Gemstone Enhancement

Other books by the author:

Gems Made by Man
Chilton Book Co, Radnor, PA (1980)

The Physics and Chemistry of Color
John Wiley and Sons, New York (1983)

Gemstone Enhancement

Heat, Irradiation, Impregnation, Dyeing, and other Treatments which alter the Appearance of Gemstones, and the Detection of such Treatments

Kurt Nassau, PhD

Butterworth–Heinemann Ltd
Halley Court, Jordan Hill, Oxford OX2 8EJ

 PART OF REED INTERNATIONAL BOOKS

OXFORD LONDON GUILDFORD BOSTON
MUNICH NEW DELHI SINGAPORE SYDNEY
TOKYO TORONTO WELLINGTON

First published 1984
Reprinted 1988, 1991

British Library Cataloguing in Publication Data
Nassau, Kurt
Gemstone enhancement
1. Precious stones 2. Gems
I. Title
549 QE392

Library of Congress Cataloging in Publication Data
Nassau, Kurt
Gemstone enhancement
Includes bibliographical references and index
1. Precious stones. I. Title
TS756.N37 1984 666.88 84—3165

ISBN 0 7506 1558 3

Printed and bound in Great Britain by
Butler & Tanner Ltd, Frome and London

To Julia

. . . . Love is not love
Which alters when it alteration finds
William Shakespeare

And to our *Apple IIe*
which has provided us with enjoyable
aspects in the writing of this book

Preface

In my work as a research scientist there are frequently occasions when I have to use the deductive-reasoning technique of the detective. This approach employs clues obtained from experimental observations to permit logical deductions to be made about what might be the nature of the processes at work behind the data. This, in turn, leads to predictions of matters as yet unobserved and the verification of such predictions then provides a strong confirmation of the understanding. My fascination with the deductive reasoning of the detective was first stimulated by reading Conan Doyle's 'Sherlock Holmes' series, G. K. Chesterton's 'Father Brown' series, and Agatha Christie's 'Miss Marple' stories, not forgetting E. C. Bentley's *Trent's Last Case* (one of the first books I read in English).

These same techniques have been useful in a number of my gemological studies, where it was more frequently the work of people, rather than the workings of nature, on which the analytical magnifying glass was focused. Examples of such studies that I undertook included the mystery of the early Geneva or 'reconstructed' ruby (natural or synthetic?), the puzzling case of Nacken's emeralds (hydrothermal growth or flux growth?), the confusing nature of deep blue Maxixe-type beryl (what was the nature of the coloration and was it natural or induced by human action?), and the processes used in the heat treatment of corundum (new discoveries or old knowledge put to new uses?).

In all of this work my aim has been to discover and explain what had been done, rather than to lead the way. As one specific example, although knowledge did exist from which I was able to deduce that heat treatment can be used to produce stable yellow or orange corundum, this was not specifically discussed in my early series of articles on corundum heat-treatments because it was not being done at the time. Only later, when such material appeared in the trade, did I include the explanation of the mechanism at work. In this book we consider all those matters which are obvious, which have been published, or which can be deduced from information that has been published.

Today, the gemologist examining, say, a blue star sapphire is faced with a multiplicity of possibilities: the stone may be natural, synthetic, or an imitation; the color may have been enhanced by heat treatment or have been added by diffusion into the surface; the asterism may have been developed by heat treatment or have been diffused into the surface; the stone may be foil-backed, it may have been oil treated or surface coated, and so on. Quite apart from the information needed to decide on whether disclosure is required, the serious gemologist will not be satisfied until he or she has determined all the information the stone is willing to reveal. This book is dedicated to the proposition that valid conclusions can be reached in such investigations only if all the possibilities are known.

Kurt Nassau, PhD
Bernardsville, NJ-07924

Acknowledgments

I have gained immeasurably from numerous discussions with many gemologists over the years, but particularly with Robert Crowningshield of the Gemological Institute of America, New York City, whose knowledge and experience are truly astonishing. Invaluable in assessing the historical aspects of treatments has been the access kindly granted by John Sinkankas to his superb library. I am also grateful to the Gemological Institute for providing access to their library and to many photographs, as acknowledged in the figure captions.

Others who have kindly provided photographs include: Bell Laboratories of Murray Hill, NJ; CM Manufacturing and Machine Co, Inc, of Bloomfield, NJ; Centorr Associates Inc, of Suncook, NH; Ceres Corp, of North Billerica, MA; Gem Instruments Corp, of Santa Monica, CA; Hanneman Gemological Instruments of Castro Valley, CA; Harper Electric Furnace Corp, of Lancaster, NY; IRT Corp, of San Diego, CA; Le Mont Scientific, of State College, PA; Omega Engineering, Inc, of Stamford, CT; and Thermolyne, Division of Sybron Corp, of Dubuque, IA.

For specimens for photographing I am grateful to Jerry Call of International Lapidaries, Boynton Beach, FL; Don McCrillis of Plumbago Mining Corp, Rumford, ME; Keith Proctor of Keith Proctor Fine Gems, Colorado Springs, CO; and Gerald V. Rogers of Allied Industries Research, Bangkok, Thailand.

Many gemologists and specialists in related fields have been most helpful in discussing specific items: these include: J. H. Borden, J. Call, B. Cass, A. Chin, A. J. Cohen, S. Frazier, R. E. Kane, J. Kean, P. C. Keller, C. L. Key, J. I. Koivula, B. Krashes, W. Larson, R. T. Liddicoat, Jr, D. V. Manson, D. McCrillis, H. Neiman, K. Proctor, G. V. Rogers, G. Rossman, M. M. Weiss, and M. A. Welt.

I am grateful to R. Crowningshield for having read the whole manuscript and to B. Cass, S. Frazier, A. Hanson, G. H. Lavoie, D. V. Manson, and M. M. Weiss for having read various parts. Their comments have been most helpful. The responsibility for any errors remains, of course, my own.

I am particularly indebted to my wife Julia, not only for typing the manuscript, but also for her continuing patience and support during the tedious periods of being an 'author's widow'!

Contents

Chapter 1

Introduction

Read not to contradict and confute,
nor to believe and take for granted,
nor to find talk and discourse,
but to weigh and consider.
Francis Bacon

One of the important tasks of the gemmologist is to identify treatments which may have been used to modify the color or appearance of the gemstones being examined. Omitted from this book are the techniques used in the shaping of faceted gemstones and cabochons, the drilling of beads and pearls, and the carving of decorative objects, all of which are well covered elsewhere. Although some of the simpler treatments were known in antiquity, there has been much random experimentation with gemstones, as well as a tremendous advance in the understanding of solid-state physics and chemistry in recent years, all leading to a variety of new and sophisticated treatment techniques. At the same time, long discontinued techniques are periodically resurrected. This book is written for gemologists to simplify their work: by knowing the details of the possible treatment techniques, the gemologist can more readily recognize which treatments have been applied to the gemstone under study. Collectors, dealers, jewelers, appraisers, pawnbrokers, and others interested in natural gemstones will also gain added insight from knowing the range of possible treatments.

It must be recognized that there is a complexity derived from the wide variability in the impurities and structural defects present in certain gemstones. Material from various sources may react in quite different ways to a specific treatment. Two sapphires of identical appearance, for example, may react differently to the same heat treatment: one may become a darker blue while the other becomes lighter, according to the relative amounts of the essential color-causing iron and titanium impurities present, as well as to other non-essential impurities. Similarly, a certain amount of irradiation may turn quartz from one locality an attractive, uniform rich-brown, smoky color, while that from another locality may be turned an impenetrable pitch black and a third specimen may darken in an irregular and patchy manner. Again, of three apparently identical smoky-quartz specimens heated to about 300 °C for 2 h, the first might remain unchanged, the second might lose all of its color, while the third might turn a greenish yellow color.

This variability in the composition and behavior of natural gemstones has the consequence that precise treatment conditions, such as irradiation doses or heating times and temperatures, cannot ever be given with any assurance that they apply to any given specimen. The person who would wish to apply the techniques here described to his or her own specimens must keep this variability in mind and proceed very carefully, so as not to produce changes which may be irreversible and could thus ruin the material. With a large piece of rough or a quantity of material known with certainty to be from exactly the same location, a preliminary test on a representative fragment can be helpful. Even this result cannot be trusted fully, since color or impurity banding is prevalent in some materials and may lead to variable results, even within one specimen, as can be seen from *Plate VII,* as one example.

A WARNING: *Anyone having a specimen on which he or she just cannot wait to try a treatment based on the specific descriptions of Chapter 7, is strongly urged to use restraint. It is absolutely essential to read through Chapters 3 to 5 and at least skim over Appendices A and B before trying any specific treatment here described. The alternative is a very high probability of instant disaster.*

A number of old treatments are given, sometimes verbatim, for their historical content; such a listing does not, however, constitute a recommendation. The venturesome reader should always first try any process on a small fragment of little value, even for modern processes described in exact detail which appear to have an assured outcome. There is no guarantee, expressed or implied, in any of the contents of this book.

Some treatments have been so widely and consistently used over the years that their employment is rarely mentioned; an example of this type of treatment is the conversion of green aquamarine to give blue aquamarine by heat. Other treatments are so drastic that disclosure is considered essential or even dictated by law; lack of such a disclosure could then void any transaction, as with a diamond that has been colored by irradiation. Many other treatments fall between these extremes, often because their use has only recently come into vogue and there may be, as yet, no general agreement on the necessity for disclosure.

The technical details of a number of treatment techniques were discovered during the development of gemstone synthesis; they apply to synthetics just as they do to natural gemstones, since both have essentially the same composition. The results with synthetics are more reproducible, however, since their composition is consistently uniform as unnecessary impurities are carefully avoided and concentrations are optimized for the best appearance.

Frequently, the use of a certain treatment is detected for the first time by a gemologist who notices unusual or inconsistent characteristics in a gemstone that is being examined. A subsequent search of the literature often reveals existing descriptions of the processes, sometimes many decades old. The initial use of the treatment may arise either because changes in the gemstone trade have made the treatment economically desirable or because, for the first time, someone aware of a need happens to have come across a reference to such a treatment. Treatments may enhance the appearance of a material directly by changing the color or by disguising imperfections, or indirectly by producing a resemblance to another material and thus providing a simulation or imitation.

The effect of a number of treatments can be reversed, an example being the frequently opposite effects of irradiation and heating. This can be important if, for example, repair that involves soldering has been attempted on the prongs holding a stone believed to be a ruby; when the color had been lost in three separate instances known to the author because the stone was not a ruby, but instead a tourmaline, the colour was successfully restored by irradiation. Such a 'restoration' is viewed by some as being quite different from the conversion of a low-value material to a high-value state.

For the purposes of this book, the field of gemstones is taken to include all materials covered by Webster in his *Gems: Their Sources, Description and Identification*[1]. In addition to the gemstones in the strict sense, Webster is followed herein by the inclusion of biologically based materials, such as pearl, coral, shell, amber, and ivory, as well as those materials employed for ornamental purposes, such as marble, alabaster, fluorite and the like. The use of 'synthetics' and imitations to substitute for natural gemstones has been dealt with in the author's *Gems Made by Man*[2] and is outside the range of interest here, as is the simple substitution, without treatment, of one natural gemstone for another, such as any colorless stone presented as a diamond imitation, or a red spinel or tourmaline presented as a ruby.

Following a historical account in Chapter 2 that shows the surprising antiquity of some treatments, general accounts are given in Chapter 3 of heat treatments, in Chapter 4 of irradiations, and in Chapter 5 of miscellaneous techniques (including impregnation, dyeing, coating, foils, and even synthetic overgrowth) that are used to enhance gemstones. It must be pointed out that heat and irradiation treatments usually modify the bulk of the treated material and are most often applied to the rough, while most other treatments only affect the surface or sub-surface regions and are usually performed on the preform or finished gem, be it a faceted stone, a cabochon, or a carved item; there are, of course, exceptions to any such generalizations. In Chapter 6 is given a brief discussion of the identification techniques used to reveal treatments, as well as some discussion of the criteria involved in the question of the disclosure of treatments.

In Chapter 7 the various gemstone materials are considered in alphabetical turn by groups, with a discussion of the various treatments which have been employed. Also included for each material, if relevant, is a brief consideration of the ways of distinguishing the different treated and untreated materials, the synthetic equivalent, and sometimes also a few of the imitations.

More detailed descriptions of heating techniques, irradiation sources, and so on, are given in Appendices A and B. Since the majority of treatments involve a modification of the color, an understanding of the various causes of color in gemstone materials is helpful in recognizing the possibilities of achieving a change, as well as in understanding the nature of the change. A detailed exposition of all the causes of color, including those active in gemstones, has appeared in the author's *The Physics and Chemistry of Color*[3]. A listing of units used is presented in Appendix C. In Appendix D is given a listing of US purveyors of relevant supplies and services.

For a general background on gems and gemology, apart from the specific references given throughout this book, the best single source is undoubtedly

Webster's *Gems*, particularly the fourth edition revised by Anderson[1]. On synthetic and imitation gemstones, there is the author's *Gems Made by Man*[2], and on gemology and the identification of the different types of materials and treatments there are Anderson's *Gem Testing*[4] and Liddicoat's *Handbook of Gem Identification*[5]. In a field which can change rapidly, a continuing scrutiny of *Gems and Gemology* and/or the *Journal of Gemmology* (*see* Appendix D) is essential.

References to the literature are given to enable the interested reader to pursue further particular topics. It must be recognized that the literature is vast and that a volume of practical size could not supply all possible citations; not that this would be desirable, since the multitude of early and often erroneous reports would frequently overwhelm the few recent and definite studies that solve the problems. For citations, one of the popular systems is used here, as in this example for a journal article:

R. Crowningshield, Unusual doublet, *Gems Gemol.*, **12,** 305 (1968)

Following the author and title, this indicates that the article appeared in Volume 12 of *Gems and Gemology,* on page 305, published in 1968 (in Santa Monica, CA). Other frequently used journal abbreviations include:

J. Gemm. – Journal of Gemmology (London, England)
Z. dt. Gemmol. Ges. – Zeitschrift der deutschen Gemmologischen Gesellschaft (Idar Oberstein, Germany)
Austral. Gemm. – The Australian Gemmologist (Sydney, Australia)
Amer. Min. – American Mineralogist (Washington, DC)
Lap. J. – Lapidary Journal (San Diego, CA)

Technical training is not required for the understanding of the account given herein. Chemical equations, absorption spectra, and detailed mechanisms are occasionally described for the benefit of those interested, but these aspects are not essential and can be passed over.

Finally, it must be recognized that the explanations for a number of the causes of specific colors and color alterations have changed significantly over the years. Reasonably well-established and plausible explanations are cited herein. Inevitably, a number of these will be found inadequate and, undoubtedly, in time there will be further reinterpretations.

References

1. R. Webster, *Gems: Their Sources, Descriptions and Identification,* 4th Edn, revised by B. W. Anderson, Butterworths, London (1983)
2. K. Nassau, *Gems Made by Man,* Chilton, Radnor, PA (1980)
3. K. Nassau, *The Physics and Chemistry of Color: The Fifteen Causes of Color,* Wiley, New York (1983)
4. B. W. Anderson, *Gem Testing,* 9th Edn, Butterworths, London (1980)
5. R. T. Liddicoat, Jr, *Handbook of Gem Identification,* 11th Edn, Gemological Institute of America, Santa Monica, CA (1981)

Chapter 2

The history of treatments

These gems have life in them:
their colours speak.
George Eliot

How many a thing
. . . becomes a gem!
George Meredith

In the brief account of this chapter it is possible to discuss only a few outstanding and historical accounts. Following detailed examinations of Pliny's *History* and the almost totally neglected *Stockholm Papyrus, P. Holm.,* a number of relevant books are concentrated upon and only occasionally are individual literature reports and references from other fields used to highlight specific points. It must kept in mind that books usually describe the techniques of a period from almost current (it can take several years to write and publish a book) to 30 or more years previously; this range is usually derived from an author's experience, which extends over his or her whole career; such a broad coverage of time is quite appropriate, since stones treated by older, discontinued processes usually continue turning up for many decades.

Pliny

As in so many areas of historical interest, the earliest prime source is C. Plinius Secundus (born 23 AD and died 79 AD during the eruption of Vesuvius), the busy compiler of all that was known in his time. His account, published in 37 books (*see Figure 2.1*), was based on notes which he said he made while reading more than 2000 books. Some of these books dealt with gemstone alterations:

> Moreover, I have in my library certain books by authors now living, whom I would under no circumstances name, wherein there are descriptions as to how to give the color of *smaragdus* [emerald, in part] to *crystallus* [rock crystal] and how to imitate other transparent gems: for example, how to make *sardonychus* [sardonyx] from a *sarda* [carnelian, in part sard]: in a word, to

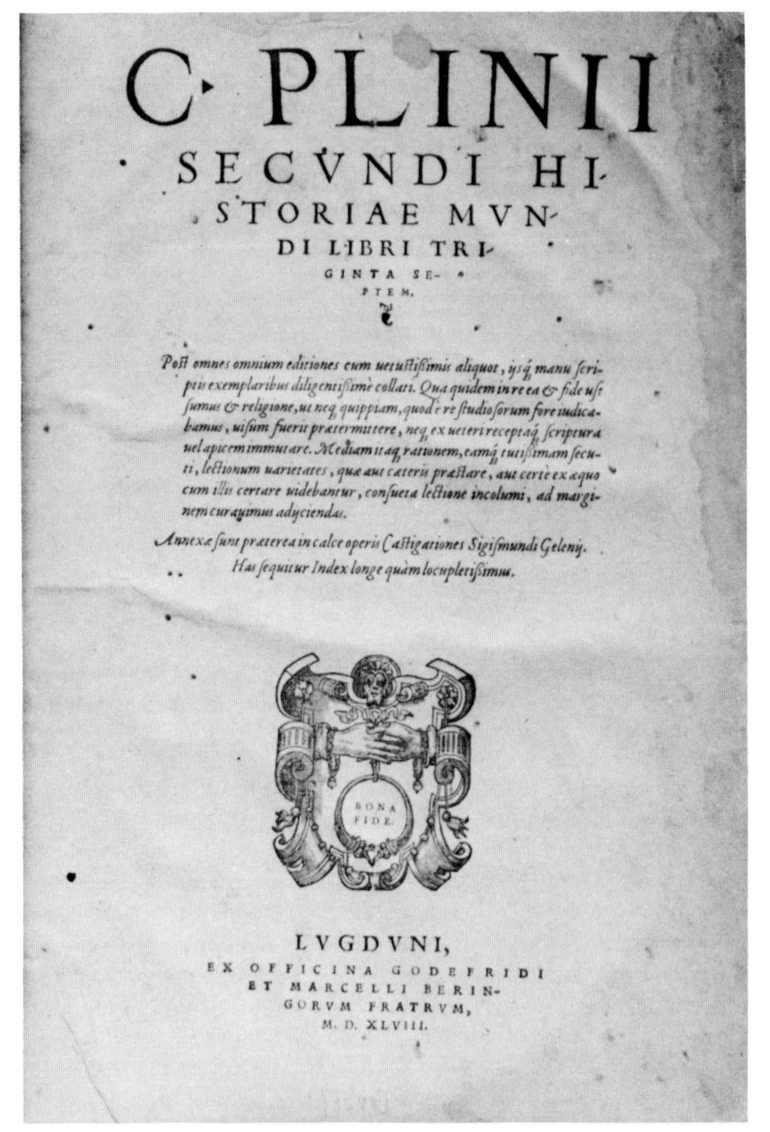
C. PLINII SECVNDI HISTORIAE MVNDI LIBRI TRIGINTA SEPTEM.

Post omnes omnium editiones cum uetustissimis aliquot, iisque manu scriptis exemplaribus diligentissimè collati. Qua quidem in re ea & fide usi sumus & religione, ut neque quippiam, quod è re studiosorum fore iudicabamus, uisum fuerit praetermittere, neque ex ueteri receptaque scriptura uel apicem immutare. Mediam itaque rationem, eamque tutissimam secuti, lectionum uarietates, quae aut caeteris praestare, aut certè ex aequo cum illis certare uidebantur, consueta lectione incolumi, ad marginem curauimus adiiciendas.

Annexae sunt praeterea in calce operis Castigationes Sigismundi Gelenii.
Has sequitur Index longe quàm locupletissimus.

LVGDVNI,
EX OFFICINA GODEFRIDI
ET MARCELLI BERINGORVM FRATRVM,
M. D. XLVIII.

Figure 2.1 Title page of a 1548 edition of the *Natural History* of C. Pliny the Second, written in the first century AD; this is the first book with descriptions of enhanced gemstones

transform one stone into another. To tell the truth, there is no fraud or deceit in the world which yields greater gain and profit than that of counterfeiting gems.
(Pliny, Book 37, Chapter 75, p.195.)[1]*

* All notes in square brackets within quotations are additions given by the author for clarification. The occasionaly quaint spellings have been carefully preserved and are not evidence of poor proofreading.

In this quotation the gemstone identifications are based on those given by Ball[1]. Apart from this elegant passage taken from the rather archaic 1601 translation by Philemon Holland which Ball used, and passing over the frequently used 1898 Bostock and Riley translation[2], only the modern Harvard University translation[3], begun by Rackham and completed by Eichholz, is hereafter cited. These editors had accessible to them more source manuscripts, as well as more sophisticated scholarship, than previous translators, including the guidance of Ball's volume in matters gemological.

Among the books that Pliny consulted may well have been one of the originals of the Greek *Papyrus Graecus Holmiensis*[4], also known as the *Stockholm Papyrus*[5], further discussed below. Unfortunately, the book *On Stones*[6] by Theophrastus (written about 350 BC), which is much more scholarly on the subject of minerals than Pliny's work, included no gemstone treatments.

Pliny discusses many gemstone enhancement techniques which are still in use today, almost 2000 years later, including foils, oiling, dyeing, and composite stones. The use of shiny metal foils to make stones appear more brilliant or to modify their color goes back at least to Minoan times (2000–1600 BC), according to Ball[1], and Pliny mentions their use on '*carbunculi* [red stones, including garnet, ruby, etc.] . . . they afford for the exercise of cunning, when craftsmen force the opaque stones to become translucent by placing foil beneath them' (Book 37, Ch 26, Vol 10, p. 243)[3]; on '*sard* [carnelian, in part sard] . . . that is backed with silver foil' (Book 37, Ch 31, Vol 10, pp. 249, 251)[3]; and with '*hyacinthus* [sapphire] and *chrysolithus* [topaz]' of quality less than the best, which 'are backed with brass foil' (Book 37, Ch 42, Vol 10, pp. 267, 269)[3].

Vinegar is used to make dull stones shiny (Book 37, Ch 26, Vol 10, p. 243)[3] and '*smaragdi* [emeralds in part] . . . . in spite of their varied colours, seem to be green by nature, since they may be improved by being steeped in oil' (Book 37, Ch 18, Vol 10, pp. 219, 221)[3]. One may assume that this refers to white or brown-appearing, badly cracked stones which become an improved green color on oiling, as is still done today. He also reported the well-known behavior of '*callaina* [turquoise] . . . . The finer specimens lose their colour if they are touched by oil, unguents or even undiluted wine' (Book 37, Ch 33, Vol 10, p. 255)[3]. In addition amber '. . . is dressed by being boiled in the fat of a suckling pig' (Book 37, Ch 11, Vol 10, p. 199).

Dyeing and staining were widely used. Even pigments made from ground-up malachite and azurite were thus improved: '*Armenian* [azurite] is a mineral that is dyed like malachite . . . .' (Book 35, Ch 28, Vol 9, p. 297)[3]. Then there was '. . . . the green called *Appian*, which counterfeits malachite; just as if there were too few spurious varieties of it already!' (Book 35, Ch 29, Vol 9, p. 297)[3]; this is a complaint still being made today of new imitations, as well as of gemstone names. In addition, '. . . . it ought to be generally known that *amber* can be tinted, as desired, with kid-suet and the root of *alkanet* [a natural dye]· Indeed it is now stained even with purple dye [Tyrian purple] . . . . *amber* plays an important part also in the making of artificial transparent gems, particularly artificial amethyst, although, as I have mentioned, it can be dyed any colour' (Book 37, Ch 12, Vol 10, pp. 201, 203)[3].

The sugar–acid process for dyeing agates and other porous stones black is apparently reported by Pliny, although the description has not always been accepted as such; this will be discussed in more detail in Chapter 5, pp. 66–70. Finally there is this passage on the making of triplets:

> . . . . men have discovered how to make genuine stones of one variety into false stones of another. For example, a *sardonyx* can be manufactured so convincingly by sticking three gems together that the artifice cannot be detected: a black stone is taken from one species, a white from another, and a vermillion-coloured stone from a third, all being excellent in their own way.
> (Book 37, Ch 75, Vol 10, p. 32.)[3]

Ball[1] cites the report back to China of the Chinese ambassador to Antioch, the capital of Roman Syria, in 97 AD: 'The articles made of rare precious stones produced in this country are sham curiosities and mostly not genuine . . . .' (p. 81) and the poet Martial, about the same time, mentions *real* sardonyx, implying the existence of the false. When one considers that almost nothing was known in Pliny's day about gemological testing, other than perhaps a very crude estimation of the hardness, it is surprising that any authentic gemstones were at all noticeable among all the fakes.

About 300 AD, Emperor Diocletian became so outraged by these activities that he ordered all books describing the fabrication of artificial gemstones to be burned[1]. It is doubtful that this edict had much effect on such activities, although it may account for the relative scarcity of surviving documents on the subject.

For the reader interested in placing these activities into the framework of the science and technology of the period, Thorndike's *History of Magic and Experimental Science*[7a] (extending from Pliny to the end of the seventeenth century), the Oxford *History of Technology*[7b], and F. S. Taylor's *The Alchemists*[7c] are detailed studies that can be recommended.

The Stockholm Papyrus: *Papyrus Graecus Holmiensis*

In 1832 the Swedish Academy (Kunglia Vitterhets-, Historie- och Antiquivitets-Akademien) received a metal box that contained fourteen numbered papyrus sheets, plus an unrelated fragment, all covered with early Greek handwriting. The gift came from Johann d'Anastasy, who was the Swedish–Norwegian Vice-Consul in Alexandria, Egypt, and an inveterate collector of early Egyptian documents. The fourteen sheets had originally been a codex, a hand-written 'book' consisting of seven folded sheets which had later been cut in half. These documents were examined by Otto Lagercrantz who published the text with a German translation and commentary in 1913[4]. Lagercrantz named it the *Papyrus Graecus Holmiensis*, abbreviated *P. Holm.,* and gave it the sub-title 'Recipes for Silver, Stones and Purple'. This papyrus is also known as the *Stockholm Papyrus*, under which title it was translated into English in 1927 by Caley[5].

From a variety of circumstances Lagercrantz deduced that this papyrus was a copy made by a scribe about 400 AD in Greek-speaking Egypt. It was probably made for the purpose of accompanying the remains of a 'chemist' in his mummy case, where it survived some fifteen centuries in excellent condition. In all probability, it was a copy of his laboratory working notes, no doubt in turn copied by a scribe from an older document. Written in uncial characters without spaces between words, translation provided serious problems, as do any texts from such early times. Although Lagercrantz was a philologist and not acquainted with gems and minerals, his commentary makes it clear that he was acquainted with relevant sources, such as Pliny and the early technical and magical writings. Neither he nor Caley, a chemist by profession, were aware of the implications to gemology of this fascinating text.

The small fragment contains a short, magical incantation of no obvious meaning. The fourteen-page main text consists of three parts. The first part deals with metals and gives nine recipes, numbers 1 to 9 in Caley's numbering, for making copper look like silver, extending silver to double its quantity, and the like. The last part, numbers 84 to 153, contains 70 recipes for the dyeing of wool and other substances. The main aim here is to imitate the very costly Tyrian purple dye, although dyeing with other colors is also described.

The longest section of 74 recipes, numbers 10 to 83, deals with the falsification of pearls and gemstones and represents the oldest extended recipe collection that deals with gems. There are an additional five recipes which were repeated when the scribe lost his place, began copying at an earlier point, recognized his mistake, and then jumped forward to the proper spot. Since this text has been generally inaccessible, more than the usual number of examples are cited herein.

There is in these recipes no attempt to duplicate anything but the color – or lack of it – of the desired gemstone. The counterfeiting is sometimes very simplistic, as in this recipe:

> *Bleaching Crystals*
> Dissolve rice in water, put the crystal in, and together with it, boil again the solution.
> (Section 56[5], pp. 187–188[4])*

Ten of the recipes deal with improving pearls or imitating them. Two are given here:

> *Cleaning a Pearl*
> When a real pearl becomes dull and dirty from use, the natives of India are accustomed to clean it in the following way. They give the pearl to a rooster to eat in the evening. In the morning they search the droppings and verify that the pearl has become clean in the crop of the bird; and, moreover, has acquired a whiteness which is not inferior to the original.
>
> *Another recipe*
> Quick lime, when it is not yet slaked in water, after having been burnt in the oven, carries hidden within it the fire; this is slaked with the milk of a dog, but that from a white bitch. Knead the lime and coat it in layers on the pearl and leave it there one day. After stripping off the lime, observe that the pearl has become white.
> (Sections 60, 61[5], p. 189[4].)

The first of these presumably relied on the acid digestive juices of the rooster to remove a thin layer of pearl; in another version (Section 25[5], p. 172[4]), the cock is cut open directly after feeding him the pearl. Versions of this technique reappear every few centuries, and are still frequently attributed to India!

When it comes to the dyeing of stones there are two separate steps involved; these steps frequently appear as separate recipes, although they are sometimes combined. First the stone, usually crystal, i.e. rock-crystal or quartz, has to be made receptive to the color. Four different Greek words are used for this preparatory step: *styphis*, meaning 'mordanting'†, but could also mean corroding

* All quotations from *P. Holm.* are the author's based on the English translation of Caley[5], using his recipe numbering, and the German translation of Lagercrantz[4], giving his pages, as well as on the Greek version given by Lagercrantz. A more detailed interpretation is being prepared for publication.

† The designation *styphis* 'mordanting' also occurs widely in the wool dyeing section of the papyrus (e.g. Section 135[5], p. 226[4]). The process there described is essentially the same as modern mordant dyeing, where an aluminum salt is precipitated on the fiber and a dye is then attached to this precipitate, as one example.

or etching (used 23 times in 14 recipes connected with stones); *areosis*, meaning 'softening', 'loosening up', or 'opening up' (used 6 times in 5 recipes); *malaksis* (used 4 times in 2 recipes) and *liosis* (used once) both meaning 'softening'. A detailed examination of the recipes (to be published) indicates that all these terms probably referred to the same process, namely cracking of the heated quartz or other stones so that the dye used in the second step could then penetrate into the cracks to produce the change in color. Mention of the heating itself is often omitted, as might be expected for something so self-evident to an expert practitioner; indeed, many of the recipes are abbreviated as in the following extreme, where three recipes are telescoped into one with only the essential ingredients given:

> *Another [Recipe for the Preparation of Green Stones]*
> Verdigris and vinegar, verdigris and oil, verdigris and calve's bile; these form emerald.
> (Section 21[5], p. 170[4].)

Sometimes the crystal is first cleaned before the preparation step:

> *Cleaning of Crystal*
> Cleaning of smoky crystal. Put it into a willow basket, place the basket into the cauldron of the [public] baths and leave the crystal there seven days. Then, when it is clean, take and mix warm lime with vinegar. Place the stone in this and let it be mordanted. Finally: color it as you wish.
> (Section 16[5], p. 164[4].)

In the next two recipes the only thing really missing is the exact temperature of the stones when they contact the liquid, so that they should crack nicely without falling apart:

> *Another [Recipe for Mordanting and Opening up Stones]*
> Put the stones into a bowl, put on it another bowl as a lid, seal the joint with clay, and let the stones be roasted under supervision for a while. Then by degrees remove the lid and pour vinegar and alum on the stones. After this color the stones with whatever dye you desire.
> (Section 54[5], p. 186[4].)
>
> *Softening Crystal*
> To soften crystal take goat's blood and dip into this the crystal which you have first heated over a gentle fire, until it is to your liking.
> (Section 36[5], p. 179[4].)

In this recipe is almost certainly found the origin of the curious fable that stones in general, and diamonds specifically, can be softened with goat's blood. Apparently, the intent of the original process was merely to 'loosen' or 'soften' the stones by cracking them for dye penetration!

The two processes, preparation and dyeing, are sometimes combined into one:

> *Mordanting for Stones*
> Let the stone stand for 30 days in putrid urine and alum. Remove the stones and insert them into soft figs or dates. These should now be treated on the coals. Blow therefore with the bellows, till the figs or the dates burn and become charred. Then seize the stone, not with the hand but with tongs, and while it is still warm place it directly into the dye bath and let it cool there. Use as many stones as you wish, however not more than 2 drachmas [each in weight]. The dye, however, should be as thick as paste.
> (Section 29[5], p. 173[4].)

Note particularly that the stone is so hot that it cannot be picked up by hand!

A wide variety of substances are used to provide the coloration itself. Some are based on copper and other metal salts, sometimes combined with the bile fluid from

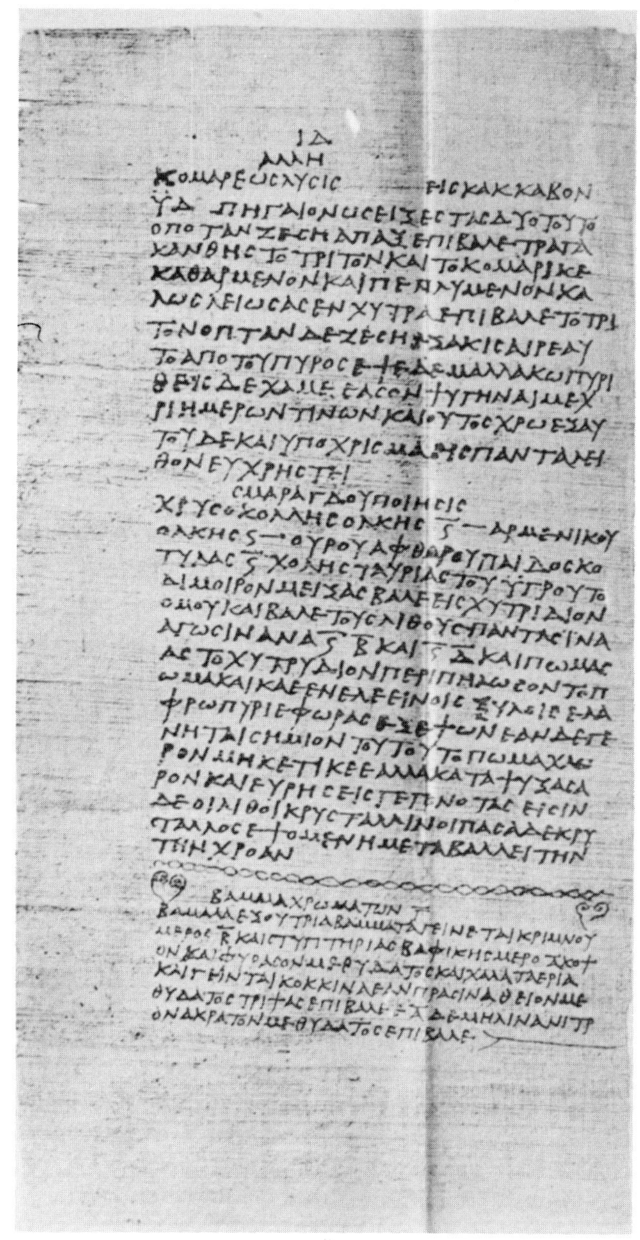

Figure 2.2 A page from *Papyrus Graecus Holmiensis*, the first manuscript to contain detailed gemstone treatment recipes, copied about 400 AD

tortoises or cattle. The following recipe is the center one of the three on the back of papyrus sheet seven, reproduced in *Figure 2.2*:

Preparation of Emerald

Mix together in a small jar 1/2 drachma of copper green [verdigris], 1/2 drachma of Armenian blue [chrysocolla], 1/2 cup of the urine of an uncorrupted youth, and 2/3 the fluid of a steer's gall. Put into this the stones, about 24 pieces weighing 1/2 obolus each. Put the lid on the jar, seal the lid all

around with clay, and heat for 6 hours over a gentle fire made of olive-wood. When there is this sign, that the lid has turned green, then heat no more, but let it cool and remove the stones. You will find that they have become emeralds. The stones are [originally] of crystal . . . .
(Section 83[5], pp. 199–200[4].)

Other colorants are based on biological substances, such as alkanet red, archil, bile, cochineal, dragon's blood, heliotrope juice, indigo, leek juice, mulberry juice, and pigeon blood. In some instances the coloring matter is added in an oily form. Here are two of these oil impregnations, one based on a copper salt and the other on leek juice:

Another [Recipe for Preparation of Emerald]
Grind scraped-off verdigris and soak it in oil one day and one night. Cook the stones in this over a gentle flame as long as desired.
(Section 77[5], p. 196[4].)

Softening of Emerald
Put hard emerald into wax for 14 days. After this time grate garlic and make a cake out of it. Take out the stone and place it into the garlic cake for 7 days. Take leeks and squeeze out the juice. Mix with the leek juice an equal amount of oil, put this into a new pot, add the stones, and boil for 3 days, until they are to your liking. The stones should be in a basket, so that they do not touch the bottom of the pot.
(Section 37[5], p. 179[4].)

Note that one is not instructed to take the stone out of the cake. Once again one suspects that the garlic cakes are baked or rather charred, just as are the figs or dates in the recipe given above, and that the stones are hot when they are dropped into the oily green leek juice to become cracked and absorb the green color at the same time.

Other oily substances used are balsam sap, Canada balsam, cedar oil, liquid pitch, resin, and wax. Oiling with colored oil is still being practiced today, as described in Chapter 5.

It is interesting to note that while the majority of the pearl recipes in *P. Holm.* deal with improvements of pearls, almost all the other gemstone recipes involve making one gemstone look like another. Only in the last-cited recipe dealing with what is now called the colored oiling of emerald (and in a variant of it, Section 72[5], p. 194[4]) does *P. Holm.* describe specifically a gemstone that is being improved as itself. Some of the other recipes that deal with non-specific 'stones' could, of course, be used for such improvements, yet the spirit of the work is clearly one of substitution rather than enhancement.

This unique manuscript represents the earliest comprehensive technical text giving explicit laboratory details. It is invaluable for the light it throws both on early chemistry and on early gemstone knowledge and techniques. Although Pliny, a few hundred years earlier, does mention some treatments, there is none of the detail which makes the processes come alive to us in *P. Holm.* To Pliny, this was mere theory; to the user of the papyrus, however, this was clearly a life's work.

The Thirteenth to the Eighteenth Centuries

For the next millenium-and-a-half, through the Dark Ages and well into the Renaissance, Pliny's work, often containing many errors from repeated copying, served as the authoritative text for matters mineralogical and gemological.

In *On Stones*, about 1260, Albertus Magnus (Albert the Great), a clergyman and emissary of Popes, gave but a single sentence of relevance to treatments. Discussing the color of precious stones in Ch 2 of Tract 2 of Book 1, he wrote '. . . . there is also found a stone having a great many colours . . . . all its colours are caused by the different substances of which its parts are composed. The same explanation holds, more or less completely, so far as the dyeing of bodies is concerned' (p. 43), a statement that seems quite straightforward when viewed in the context of precious stones and the dyeing processes as revealed by Pliny and others, but which puzzled the translator[8].

In 1502 was published *The Mirror of Stones*, dedicated to Cesare Borgia, a fascinating book by Camillus Leonardus, MD (also known as Camillo Leonardi or

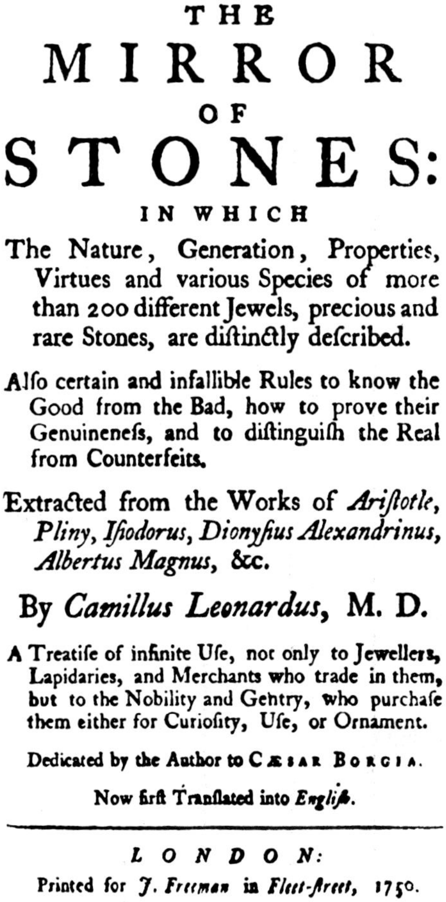

THE

MIRROR

OF

STONES:

IN WHICH

The Nature, Generation, Properties, Virtues and various Species of more than 200 different Jewels, precious and rare Stones, are diſtinctly deſcribed.

Alſo certain and infallible Rules to know the Good from the Bad, how to prove their Genuineneſs, and to diſtinguiſh the Real from Counterfeits.

Extracted from the Works of *Ariſtotle*, *Pliny*, *Iſiodorus*, *Dionyſius Alexandrinus*, *Albertus Magnus*, &c.

By *Camillus Leonardus*, M. D.

A Treatiſe of infinite Uſe, not only to Jewellers, Lapidaries, and Merchants who trade in them, but to the Nobility and Gentry, who purchaſe them either for Curioſity, Uſe, or Ornament.

Dedicated by the Author to CÆSAR BORGIA.

Now firſt Tranſlated into *Engliſh*.

LONDON:

Printed for *J. Freeman* in *Fleet-ſtreet*, 1750.

Figure 2.3 The title page of the Leonardus work on gemstones, published in 1502 and translated into English in 1750

Lunardi), a physician and astrologer of Pesaro, Italy. The title page, shown in *Figure 2.3*, acknowledges his sources, including Pliny. Chapter IX of Book 1 is given in its entirety in the 1750 translation[9], as follows:

How to know whether Jewels are natural or artificial

Since these Times abound with Counterfeits in every Thing, but especially in the Jewelling Art in regard to their Value; and as there are few unless such as have been long practis'd in them, can judge of them, especially when they are cemented together; and that we may not be deceived by these, nor leave any Thing untouch'd relating to the Subject, we shall close the First Book with a few Things upon this Head. We say then, that these deceitful Artists in Stones have many Ways of Imposition. As first, when they make Stones of a less Value, and of a particular Species, appear of another Species and consequently of a higher Price; as the *Balasius* of the *Amethist*, which they perforate, and fill the Hole with a Tincture, or bind it with a Ring, or more subtilly, when they work up the Leaves of the *Balasius*, either with *Citron Sapphire* or *Beril*, into the Form of Diamonds, and by adding a Tincture to bind them, sell them for true Diamond. Or, very often they fabricate the upper Superficies of the *Granate*, and the lower of *Chrystal*, which they cement with a certain Glew or Tincture; so that when they are set in Rings they appear like *Rubies*. And many other Deceptions may be effected out of divers and various Stones, which are all known to the Skilful. Therefore, when there is a Suspicion, the Jewels are to be taken out of the Rings, and by what we have farther to say in the Second Book, we may easily judge of them. A Deception may happen in another Manner; as when they make the Form and Colour of a true Stone from one that is not true. And this Deception is made from many Things, and chiefly from smelted Glass, or of a certain Stone, with which our Glass-makers whiten their Vessels, by adding divers permanent Colours to the Fire, as the Potters know; and as I have often seen *Emeralds*, far from bad ones, at least for Use, made out of these Stones. These counterfeit Stones may be known many Ways, as first by the File, to which all false Stones give Way, and all natural ones are Proof against, except the *Emerald* and the Western *Topaz*, as we shall shew in the Second Book; and therefore these Falsifiers chuse to work upon these which give way to the File, because they cannot be prov'd by it. The second Way to prove them is by the Aspect; for such as are natural, the more they are look'd at, the more the Eye is delighted with them; and when they are held up to the Light of the Candle, they shine and look fulgent. Whereas the Non-naturals, or artificial, the more they are beheld, the more the Sight is wearied and displeas'd, and their Splendor seems continually decaying, especially when they are oppos'd to the Light of a Candle. They are also known by their weight when they are out of the Rings; for those which are natural are ponderous, except the Emerald, but the Artificial are light. There is one Proof yet remaining, which is infallible, and is prefeable to all the rest; namely, that the Artificial do not resist the Fire, but are liquefied in it, and lose their Colour and Form when they are dissolved by the Fierceness of the Fire; and it is impossible but that in some Parts of them, some Points like small Bubbles must be seen in their Substance, produc'd by the igneous Heat, and will discover the Disproportion in their Composition, and their Difference from Nature in true Stones. Such false Stones may likewise be compounded of other Things than of Glass, namely, of many Minerals; as of Salt, Copperas, Metals, and other Things, and as I have seen, and is allowed by many learned Men, especially by Brother *Bonaventure* in the Second Book of his Dictionary of Words, that the Knowledge of Stones, and their Species, is acquired by great Experience, and from continual Uses, as they well know who employ themselves in this Kind of Exercise. And here we shall conclude this first Book.

The range of treatments here alluded to includes an astonishing variety of colorations and assembled stones. His testing techniques, especially the heating in the fire, no doubt destroyed many genuine gemstones. Particularly noteworthy is his astute observation of the presence of small bubbles in imitations made of glass.

Next the accounts of two master craftsmen in metallurgy and related arts are considered: the *Pirotechnia* of Vannoccio Biringuccio[10], printed in 1540 in Venice, and the *Treatise on Goldsmithing*[11] of Benvenuto Cellini, published in Florence in 1568 and translated into English in 1898. Both mention the use of colored foils placed behind gemstones, both discuss the use of a black backing or coating on diamond, and both mention the heat treatment of a sapphire to turn it colorless. Biringuccio says of sapphire: '. . . . The best are the oriental ones. It can be made to lose its color by keeping it in molten gold over a fire for twenty-four hours. With these baths they disguise it in the form of a diamond and try to deceive people' (Book 2, Ch 13, p. 125). Cellini puts it thus: 'There are certain sapphires, which the ingenuity of man can turn white, by putting them in a crucible in which gold is to be melted ['Nel quale sia dell'oro che s'abbia a struggere'], and if not at the first heating, then at the second or third' (Ch 10, pp. 40–41).

Cellini gives highly detailed accounts of various treatment processes, particularly the use of shiny foils including colored ones (Chs 5 to 7, pp. 24–29) and even colored cloth (Ch 6, p. 25), behind the gemstone in the cavity of the setting. In Italy at that time such activities were permitted with all gemstones, but tinting colored gemstones, such as emerald, ruby, and sapphire, was strictly forbidden by law. Curiously enough, the tinting of diamonds was permitted. There are details of a large diamond given by Emperor Charles V to Pope Paul, who gave it to Cellini to make an elaborate setting and to tint it. He used a clear undercoat of carefully selected pieces of gum mastic followed by a smoky layer consisting of a mixture of freshly prepared soot, selected gum mastic, freshly pressed linseed oil, almond oil, and turpentine, and so almost doubled the value of the stone, from 12 000 scudi to 20 000 scudi (Chs 8 and 9, pp. 31–39).

Cellini also reports that others used the blue dye indigo for tinting diamond, particularly yellow ones, which: '. . . . they make green, hence the yellow diamond with the blue tint made an admirable water; and, if it be well applied, it becomes one colour, neither yellow as heretofore nor blue owing to the virtue of the tint, but a variation, in truth, most gracious to the eye' (Ch 9, p. 36). Here he uses the quality term 'water of the diamond' in the sense of achieving the most desirable pale blue–green or smoky colors, not in the usually attributed colorless sense. Could this perhaps have been the origin of this designation? Another technique was to use a black backing on diamond. In the words of Biringuccio: 'diamond . . . . is . . . . harder and much more lustrous and transparent than any other thing. If the skin of its earthiness is cleaned with art and then it is given its polish, it becomes very brilliant when a lustrous black color is placed underneath.' (Book 2, Ch 13, p. 122.)

Illegal falsification was also achieved by the coating of pale stones according to Cellini '. . . . I once saw a ruby of this nature falsified ever so cleverly by one of these cheats. He had done it by smearing its base with dragon's blood* . . . you would gladly have given 100 scudi for it; but without the colour it wouldn't have fetched 10 . . . the colour looked so fine and the stone seemed so cunningly set, that no one unless very careful, would have spotted it' (Ch 6, p. 26). Doublets were also widely made:

> I mind me also of having seen rubies and emeralds made double, like red & green crystals, stuck together, the stone being in two pieces, and their usual name is 'doppie' or doublets. These false stones are made in Milan, set in silver, and are much in vogue among the peasant folk; the ingenuity of man has devised them to satisfy the wants of these poor people when they wish to make presents at weddings, ceremonies, and so forth, to their wives, who of course don't know any difference between the real and the sham stone, & whom the little deceit makes very happy. Certain avaricious men however, have taken advantage of a form of industry, made partly for a useful, and partly for a good end, & have very cunningly turned it to great evil. For instance, they have taken a thin piece of Indian ruby, and with very cunning setting have twisted and pieced together beneath it bits of glass which they then fixed in this manner in an elaborate & beautiful setting for the ring or whatever it was. And these they have subsequently sold for a good and first-class stone. And forasmuch as I don't tell you anything unless I can illustrate it by some practical example, I'll just mention that there was in my time a Milanese jeweller who had so cleverly counterfeited an emerald in this way that he sold it for a genuine stone and got 9000 golden scudi for it. And this all happened because the purchaser – who was no less a person than the King of England – put rather more faith in the jeweller than he ought to have done. The fraud was not found out till several years after.
> (Ch 6, p. 27.)

* Dragon's blood is a natural vegetable dye but the name was also sometimes applied to the red lead compound *minium* Pb_3O_4 according to Agricola[12].

In his *Textbook of Mineralogy*[13], published in 1556, Georgius Agricola discusses metal foils and notes:

> If one suspects that a gem has been embellished and enlightened with foil and wishes to know the true color of the gem, he must take the stone out of the ring and remove the coloring substance. Since fraud begins with a single act and once begun is hard to stop, I shall mention a few of the many ways in which gems are falsified as well as a few ways in which true gems can be distinguished from the false so that anyone may detect them and thus protect himself against fraud.
> (Book 6, p. 115.)

He discusses doublets (triplets in modern terminology) made of glass, of quartz, or of garnet and quartz, with a layer of a dye; there are also doublets that consist of a diamond top with a base of quartz, corundum, or beryl (emerald). Then there are 'filled' gemstones: 'Certain amethysts are perforated and filled with minium or are deeply engraved and thin sheets of foil cemented beneath them so that they may be passed as carbunculi' (Book 6, p. 116). Finally there is dyeing: 'The light-coloured or feminine sapphires can be darkened by dyeing. A king of Egypt was the first to dye this stone. Quartz and glass are also dyed to imitate sapphire . . . .' (Book 6, p. 130).

Figure 2.4 An assay furnace from Agricola's 1556 book on metallurgy

There is no discussion of gemstone treatments in Agricola's *De Re Metallica*[13], but there are many illustrations of the types of furnaces used in his time in the field of metallurgy, as shown in *Figures 2.4* and *2.5* and elsewhere in this book; these same types of furnaces were undoubtedly also used for heat-treating gemstones.

In the seventeenth century, *Gemmarum et Lapidum Historia*, 1609, was produced by Boetius de Boot, a physician of Bruges; this appeared in a number of different forms over the next 40-plus years[14]. In the French translation of 1644[15] there is discussion in Chs 20 to 22 of the decolorizing by heat of sapphire, topaz, amethyst, etc., to produce diamond imitations; the dyeing of stones, mostly with metal compounds (the use of gum mastic is considered to be 'trivial and vulgar'); an extended discussion of metal foils; and an obscure description on how to harden gemstones. Several of the techniques are attributed to Baptiste de la Porte or de

Figure 2.5 A glass furnace from Agricola's 1556 book on metallurgy

Porta, presumably derived from personal contact or from an earlier version of the item described below, which was published at a later date.

Next there is John Baptista Porta with his *Natural Magic*[16] published in London in 1658. Book 6 'Of counterfeiting Precious Sones' (sic) covers a variety of topics, including the making of colored glass imitations and various treatments. Chapter 5, 'How Gems are coloured', is devoted to dyeing sapphire, amethyst, topaz, chrysolite, and emerald. In a detailed passage he teaches how:

> *To turn a Saphire into a Diamond*
> This stone, as all others, being put in the fire, loseth his colour: For the force of the fire maketh the colour fade. Many do it several ways: for some melt gold, and put the Saphire in the middle of it; others put it on a plate of iron, and set it in the middle of the fornace of reverberation; others bury it in the middle of a heap of iron dust. I am wont to do it a safer way, thus: I fill an earthen pot with unkill'd lime, in the middle of which I place my Saphire, and cover it over with coals, which being kindled, I stop the bellows from blowing, for they will make it flie in pieces. When I think it changed, I take a care that the fire may go out it self [that is not pouring water on it]: and then taking out the stone, I see whether it hath contracted a sufficient whiteness; if it have, I put it again in its former place, and let it cool with the fire; if not, I cover it again, often looking on it, until the force of the fire have consumed all the colour, which it will do in five or six hours; if you find that the colour be not quite vanished, do again as before, until it be perfect white. You must be very diligent, that the fire do heat by degrees, and also cool; for it often happeneth, that sudden cold doth either make it congeal, or flie in pieces. All other stones lose their colour, like the Saphire; some sooner, some later, according to their hardness. For the Amethist you must use but a soft and gentle fire; for a vehement one will over-harden it, and turn it to dust. This is the art we use, to turn other precious stones into Diamonds, which being cut in the middle, and coloured, maketh another kind of adulterating Gems; which by this experiment we will make known: And it is
>
> *How to make a stone white on one side, and red or blew on the other*
> I have seen precious stones thus made, and in great esteem with great persons, being of two

> colours: on one side a Saphire, and on the other a Diamond, and so of divers colours. Which may be done after this manner: For example, we would have a Saphire should be white on one side, and blew on the other; or should be white on one side, and red on the other: thus it may be done. Plaister up that side which you would have red or blew, with chalk, and let it be dryed; then commit it to the fire, those ways we spoke of before, and the naked side will lose the colour and turn white, that it will seem a miracle of Nature, to those that know not by how slight an art it may be done.
> (pp. 183–184.)

The significance of this passage is discussed in Chapter 3 under Heat Treatments, p. 37.

Porta's Chs 10 and 11 deal with 'Of leaves of Metal to be put under Gems' and 'How leaves of Metals are to be polished'. Porta's book appears to have been one of the first of a continuing series of recipe books. Such books are intended to give the general public detailed instruction. They are particularly valuable in our investigation: since they are written for non-professionals, the directions tend to be much more detailed than are texts intended for the professional reader.

Robert Boyle, the famous chemist, published in 1672 *An Essay About the Origine and Virtues of Gems*[17]; this was the first work on gems written by a professional scientist who based his deductions on his own experiments and observations. The essay deals mostly with the origin of gemstones, but several treatments are mentioned in passing. Most stones lose their color on being heated, but some agates '. . . . where there ran little Veins, that I ghess to be of a Metallic Nature, there, I say, the Colour was not destroyed, but chang'd, and the Veins of Pigment thus colour'd acquir'd a deep redness, which they will retain' (p. 29). There is also a rather obscure mention of a 'tincture' which beautified a garnet and other gems by merely soaking them in the cold (pp. 32–33). The most fascinating passage is a detailed account of how to make dyed, crackled quartz, a process already described in *P. Holm.*[4] 1300 years previously, and still used today:

> I have sometimes taken pleasure to heat a piece of Christal [quartz] red hot in a Crucible and then quench it in Cold Water: For ev'n when the parts did not fly or fall asunder, but the Body retain'd its former shape, the multitude of little Cracks that were by this operation produc'd in it, made it quite loose its transparency and appear a White Body. In making which experiment, the multitude of produc'd flaws may be pretty well discover'd to the incredulous, if, as I have sometimes done, the ignited Chrystal be warily and dextrously quench'd not in Water but in a very deep solution of Cochaneel [the red dye cochineal] made with Spirit of Wine, in which operation, if it be well performed, (but not otherwise,) enough of the red Particles of the solution will get into the cracks of the Chrystal, to give it a Pleasing Colour.
> (pp. 26–27.)

A superb recipe book is the *Secretos*[18] of Don Bernardo Monton, published in Madrid in 1760. The title page is shown in *Figure 2.6*. Here there are instructions on how to dye garnets to make them look like real rubies, to soften ivory or make old ivory look like new, to bleach diamonds, to harden stones, to whiten pearls, alabaster, and marble, to color marble, and so on. Some of these seem not unreasonable, such as:

> *To make Garnets look like real Rubies*
> Take one oz. of aqua fortis [nitric acid] in which one has dissolved 1 oz. of steel filings: when dissolved add one oz of fine copper and one oz. of mercury; put the garnets in this solution and put it over the fire of a burner with three or four wicks: let it boil a little more than one hour: then take out the garnets, dry them, and wash them with clear water, dry them again, and they will look like real rubies: make sure that the garnets are transparent and of good quality.
> (pp. 5–6, author's translation.)

SECRETOS
DE ARTES LIBERALES,
Y MECANICAS,
RECOPILADOS, Y TRADUCIDOS
de varios, y ſelectos Autores,
QUE TRATAN DE PHISICA,
Pintura, Arquitectura, Optica, Chimica, Doradura, y Charoles, con otras varias curioſidades ingenioſas.
SU AUTOR
EL LIC. DON BERNARDO MONTON.

CON LICENCIA. En Madrid, en la Imprenta, y Libreria de Joſeph Garcia Lanza, Plazuela del Angel, donde ſe hallará. Año 1760.

Figure 2.6 The title page of the Monton 1760 book, which contains many gemstone treatment recipes

Others are a little less promising, such as:

> *How to bleach diamonds*
> Take barley meal and verdigris in equal amounts, then take a red-hot magnet stone and quench it in strong vinegar: repeat this eight to ten times: then take the vinegar, barley meal and verdigris and make a paste into which you put the diamond, then evaporate over a low heat and then raise the temperature for four hours.
> (pp. 45–46, author's translation.)

And some are totally obscure:

> *A very noble secret to harden stones, to make them look like diamonds*
> In a retort you place alum crystals and distill them and then again put water on them three to four times, always distilling: then take this water and put it in a matras casserole with the carved crystal stones: put this over hot ashes and next under the casserole you place a burner of three or four wicks for thirty days, everything in your oven being well regulated.
> (p. 115, author's translation.)

There are some modern gem-treatment recipes which make no more sense than this!

During this period there was a growing interest in the field, that paralleled the rise of commerce and the spread of wealth, which in turn provided a wider market for gemstones, as well as for their enhancement.

The Nineteenth Century

The nineteenth century is an era which produced gemology textbooks proper, such as those of Mawe in 1813[19], Feuchtwangler in 1838[20] and 1859[21], Kluge in

1860[22], and the series of monumental texts of King starting in 1860[23–26]. Mawe[19] reports of sapphire:

> The pale varieties when exposed to a strong fire become entirely colourless without undergoing any other alteration; after this, when cut and polished, they have been often sold as Diamonds; on this, (somewhat fraudulent) account, they bear considerable value.
> (p. 71.)

This slightly confusing statement implies that in those days pale blue sapphires had considerable value because they could be heat treated to become colorless and serve as diamond imitations; ironically, today they again have considerable value, but this time because they can be heat treated to become a dark blue color.

Mawe[19] also reports, on p. 67, that ruby is not changed by the most intense heat. Feuchtwangler[20] is much more sophisticated: 'If the red Sapphire (Ruby) is exposed to a great heat, it becomes green, but when cold, returns to its original colour' (p. 69); he also says[20]: 'Many Sapphires may be deprived of their specks by a careful calcination in a curcible [sic] filled with ashes or clay, and they assume then a more agreeable and purer colour and greater transparency.' (p. 72.)

Kluge[22] mentions the heating of topaz to turn it pink, the loss of color of sapphire, zircon, amethyst, etc., when strongly heated, usually buried in iron filings or burned lime. Using a mixture of iron filings and sand, a careful but less intense heating can remove spots and defects even in ruby. A pearl which has lost its luster may be fed to a chicken which is slaughtered after one minute; compare this with the account in *P. Holm.*[4], some 1500 years earlier! A secret process for tinting pearls yellow is mentioned, as are a large variety of other techniques for treating them, including baking them inside a loaf of bread dough! There is a discussion of the improvement of amber and an extended treatment of the dyeing of agates, based on the techniques then recently discovered in 'Oberstein and Idar' and further discussed herein in Chapter 5, pp. 60–70.

There is one last recipe book worthy of mention: *Dick's Encyclopedia of Practical Receipts and Processes*[27], published in the US about 1880. Among the 6422 recipes given, there are instructions for dyeing or bleaching ivory, alabaster, marble, etc. (recipes Nos. 1982 to 2047), a cement for amber (2176), as well as a variety of shiny and colored foils with instructions for their use (2448 to 2459). One typical quotation will suffice:

> *2459. To Make an Imitation Diamond more Brilliant*
> Cover the inside of the socket in which the stone or paste is to be set with tin foil, by means of a little stiff gum or size; when dry, polish the surface, heat the socket, fill it with warm quicksilver, let it rest for 2 or 3 minutes, after which pour it out and gently fit in the stone; lastly, well close the work round the stone, to prevent the alloy being shaken out. Or: Coat the bottom of the stone with a film of real silver, by precipitating it from a solution of the nitrate in spirits of ammonia, by means of the oils of cassia and cloves. (*See* Silvering Glass.) Both these methods vastly increase the brilliancy both of real and factitious gems.
> (p.230.)

A certain uniformity sets in, starting about the middle of the nineteenth century; as gemology turned into a science, authors studied each other's books and soon there arose a set of generally agreed facts. As a result, it is not useful to consider, on a one-by-one basis, the gemological texts beyond the middle of the nineteenth century. The works of King[26] and Bauer[28] are representative of the level of knowledge and understanding of gemstone treatments for this period and well into the twentieth century, until the period when the discovery of irradiation provided a new approach to gemstone enhancement.

The 'Precious Stones' of M. Bauer

Max Bauer's *Edelsteinkunde* of 1896, translated and published in English as *Precious Stones*[28] in 1904 and reprinted in 1968, gives a good presentation of the state of gemstone enhancement at the beginning of this century and is therefore worthy of examination in some detail. Bauer's work is so extensive that it remains today, more than eighty years later, an invaluable reference book for the dedicated gemologist.

The loss (or change) of color with heat is described by Bauer for smoky quartz, amethyst (to yellow), zircon, and topaz (to pink) (p. 58); this plus burning, the heating of carnelian to become red and the dyeing of cracked rock-crystal are further discussed (pp. 87–88). Foils and other colored backings are covered, as well as pigments applied to the girdle of a stone (pp. 90–91). Doublets of varying types, including hollow ones (pp. 96–97), are discussed, as is the heating of some green diamonds to convert them to yellow, violet, or brown, while some brown ones become greyish (p. 136). Diamond-topped doublets and blue-tinted diamonds are described (p. 260).

Sapphire may lose its color with heat and may be backed with a blue–silver foil (pp. 266, 283–285), while unevenly colored ruby may have the color evened by a heat treatment (p. 266). Foils are used on aquamarine (p. 319). Some yellow topaz may be burned to pink, while some yellow or pale blue stones may lose their color on exposure to sunlight (p. 332). Zircon can be made lighter in color or completely colorless by heating, and some fade when exposed to light (p. 342). Opal can be oiled and backed by foil, etc., or treated to become black by an unknown process (pp. 376–377, 379), while hydrophane opal may be soaked in water, oil, wax, or colored liquids (p. 387). Turquoise may bleach, fade, or alter in color; ammonia or grease may restore the color and dyeing may also be used (p. 391). Ivory is dyed to imitate bone turquoise (p. 402). Foil, either colorless or green, may be used on olivine (p. 406). The color of lapis lazuli may be improved by heating (p. 439).

The luster of colorless quartz can be slightly increased by heating (p. 475); smoky quartz turns yellow and then colorless on heating (p. 480). Amethyst may be burned to give citrine (pp. 482, 486) and rock crystal may be dyed to look like amethyst (p. 486). The yellow color of tiger's-eye can be removed with acid (p. 495). Chrysoprase may lose its green color on heating or on exposure to sunlight; this may be restored with water or with dyes, which latter can also be used on chalcedony, sard, agate, etc., to imitate chrysoprase (p. 497) as well as to obtain other colors (pp. 505, 510, 522–524). Yellow, brown, gray, and other colors of sard, carnelian, etc., may be heated to obtain red or other colors (pp. 509, 524).

The color of fluorite is altered or lost on being heated and this material may also be resin impregnated (pp. 529–530). Amber is clarified and softened on being heated with rapeseed or linseed oil, a process that can also produce sun-spangles (pp. 538–539, 546); then there is pressed amber or 'ambroid', made by heating and pressing at the same time (pp. 552–553). Amber darkens even in the dark (p. 539). The lost luster of pearls can only rarely be restored by peeling (p. 589). The red or black color of coral is lost on heating (p. 605).

This brief resumé is based on a rapid scan. Note that foils for the backing of stones had become relatively unimportant by this time, based on the widespread use of quality faceting employing angles giving optimum light reflection, as described here in Chapter 5. It is curious that there is no mention of the oiling of emeralds or the heat treatment of aquamarines. The relative scarcity of oiling, as well as of dyeing, should also be noted, except for widespread dyeing used on the microcrystalline quartz family.

The Twentieth Century

X-Rays were discovered by W. K. Röntgen in 1895, radioactivity in uranium by H. Becquerel the following year, and gamma rays by P.-U. Villard in 1900. Within ten years there began a veritable flood of studies of the effect of these various rays and particles on gemstone materials that has not yet come to a complete halt.

Grossman and Neuburger in their 1918 book[29] described how it all began when the Curies noted that radium, which they discovered in 1898, produced a bluish color in the glass tube in which it was stored – presumably the color now sometimes designated as desert-amethyst glass[30]. Next, F. Bordas in the Laboratory of the Collège de France in Paris buried some corundums for one month in a radium salt and obtained, in colorless corundum, a yellow color, while a blue corundum turned an emerald green, and so on; he reported that the jeweller who had sold the stones to Borads for two Francs per carat now offered him 45 Francs per carat for them.

Already in his 1909 book on diamonds[31], W. Crookes reported the production of a green color on exposing colorless diamonds to radium, resulting in 'fancy stones' having an increased value. He found that the stones were radioactive, and that this activity was not lost on heating the stones. Also in 1909, Doelter[32], a very active worker in this field, pointed out that some of the irradiation-induced colors would fade on exposure to light. He found that colorless fluorite irradiated to become violet faded completely in light, that pink topaz irradiated to orange faded partly, but that irradiation-colored diamond did not fade. Heating accelerated the fading: 100 °C was enough to bleach fluorite rapidly and 300 °C did the same for the yellow color produced in blue corundum, as well as for rose quartz, citrine, and smoky quartz.

A good summary of early work and the results achieved to that date is given by Michel in his 1926 book on artificial gemstones[33]. The many results presented include expected ones, as well as some no longer accepted: the latter no doubt, at least in part, because of misidentification of some of the gemstone materials used in the experiments. Based on a report by Michel published in translation in 1927[34], the significant changes known at that time were the development of yellow, green, or blue in diamond, brown in the quartzes, yellow in corundum, red or yellow in tourmaline, orange in topaz, and green in kunzite.

As more intense sources of radiation became available, there were further reports covering ever wider ranges of material, including the two by Pough and Rogers[35] and Pough[36]. As one example of the delayed use of such information, the Pough report[36] mentions the development of a blue color in some topaz on irradiation. This author observed one irradiated and heated specimen having a blue color in a 1973 study of the effect, on many quartz samples, of irradiation followed by heating; this particular effect was found to have occurred in a specimen of topaz accidentally mixed in with the quartz. When this was reported[37], it was viewed in the gemstone trade as a new discovery and was seen as explaining the origin of significant amounts of blue topaz then appearing on the market, without any new mines having been discovered! By now this activity has turned into a full-fledged industry with a flood of products of a blue color even deeper than that of the best, previously known natural topaz.

In addition to his discussion of synthetics and irradiations, Michel in his 1926 book[33] gave a fairly detailed description of various kinds of doublets (pp. 371–376), an outline of color-changing heat treatments (pp. 376–378), including brown topaz to pink, zircon to colorless or blue, and amethyst to yellow. He also discussed the dyeing of agate, chalcedony, opal and the like, as well as the production of dendritic patterns by the use of a silver nitrate solution, the production of crackled quartz, and but a brief mention of foils (pp. 378–381). There is also a discussion of the bleaching and dyeing of pearls, subjects not mentioned by Bauer[28] less than 30 years previously.

The only additional irradiation developments in recent years have been the perfection of the diamond processes to control the production of attractive shades of yellow, green, blue, and sometimes pink, and the irradiation of beryl to produce the fugitive Maxixe-type, deep-blue color center. Details of these subjects are covered in Chapter 7. Aside from these irradiations, the major developments of the last 50 years have been laser drilling and laser engraving of diamonds, the as yet unknown process believed to be used to produce a lavender color in jade, the bleaching of black coral and, most recently, the spectacular expansion of corundum heating, which produces yellow, orange, or deep blue stones from low quality rough, and even the use of diffusion to add color and/or asterism. These too, are covered in Chapters 3 and 7.

All the time there are, of course, improvements in some of these processes as well as periodic resurgences of some of the oldest enhancement techniques which have been almost totally forgotten and so may catch the unwary gemologist by surprise.

References

1. S. H. Ball, *A Roman Book on Precious Stones,* Gemological Institute of America, Los Angeles, CA (1950)
2. J. Bostock and H. T. Riley, *The Natural History of Pliny,* George Bell & Sons, London, 5 vols (1893–1898)

3. H. Rackham, *Pliny: Natural History,* Harvard University Press, Cambridge, MA, Vols 1–5 and 9 (1938–1952); continued by W. H. Jones, Vols 6–8 (1951–1963); completed by D. E. Eichholz, Vol 10 (1962)
4. O. Lagercrantz, *Papyrus Graecus Holmiensis (P. Holm.), Recepte fur Silber, Steine and Purpur,* Uppsala (1913)
5. E. R. Caley, The Stockholm Papyrus, *J. chem. Educ.,* **4,** 979–1002 (1927)
6. E. R. Caley and J. F. C. Richards, *Theophrastus on Stones,* Ohio State University, Columbus (1956)
7a. L. Thorndike, *A History of Magic and Experimental Science,* Columbia University Press, New York, 8 Vols (1923–1958)
7b. C. Singer, E. J. Holford, and A. R. Hall (Eds), *A History of Technology,* Clarendon Press, Oxford, 7 Vols (1954–1978)
7c. F. S. Taylor, *The Alchemists, Founders of Modern Chemistry,* H. Schuman, New York (1936)
8. D. Wyckoff, *Albert Magnus: Book of Minerals,* p. 43, Clarendon Press, Oxford (1967)
9. C. Leonardus, *The Mirror of Stones,* J. Freeman, London (1750)
10. V. Biringuccio, *Pirotechnia,* MIT Press, Cambridge, MA (1942; paperback edition 1966)
11. B. Cellini, *The Treatises of Benvenuto Cellini on Goldsmithing and Sculpture,* Dover Publications, New York (1967)
12. G. Agricola, *De Natura Fossilium, Special Paper 65,* Geological Society of America, New York (1955)
13. G. Agricola, *De Re Metallica,* Basel (1556)
14. A. B. de Boot, *Gemmarum et Lapidum Historia* (1609), cited in J. R. Partington, *A History of Chemistry,* Vol 2, pp. 101–102, Macmillan, London (1961)
15. A. B. de Boot, *Le Parfaict Ioaillier ou Histoire des Pierreries,* Lyon (1644)
16. J. B. Porta, *Natural Magic,* Basic Books, New York (1957)
17. R. Boyle, *An Essay About the Origine and Virtues of Gems,* Hafner Pub Co, New York (1972)
18. B. Monton, *Secretos de Artes Liberales y Mecanicas,* Madrid (1760)
19. J. Mawe, *A Treatise on Diamonds and Precious Stones,* London (1813)
20. L. Feuchtwangler, *A. Treatise on Gems,* New York (1838)
21. L. Feuchtwangler, *A Popular Treatise on Gems,* New York (1859)
22. K. E. Kluge, *Handbuch der Edelsteinkunde,* Brockhaus, Leipzig (1860)
23. C. W. King, *Antique Gems: Their Origin, Use, and Value,* London (1860)
24. C. W. King, *The Natural History, Ancient and Modern, of Precious Stones,* London (1865)
25. C. W. King, *The Natural History of Gems or Decorative Stones,* London (1867)
26. C. W. King, *The Natural History of Precious Stones and of the Precious Metals,* London (1883)
27. W. B. Dick, *Dick's Encyclopedia of Practical Receipts and Processes, or How They Did It in the 1870s,* Funk and Wagnall, New York (1974)
28. M. Bauer, *Precious Stones,* Dover Publications, New York, 2 Vols (1968)
29. H. Grossman and A. Neuburger, *Die Synthetischen Edelsteine,* Berlin (1918)
30. K. Nassau, *The Physics and Chemistry of Color: The Fifteen Causes of Color,* Wiley, New York (1983)
31. W. Crookes, *Diamonds,* London (1909)
32. C. Doelter, Ueber die Stabilität der durch Radium erhaltene Farben der Mineralien, *Zbl. Miner. Geol. Paläont.,* **8,** 232 (1909)
33. H. Michel, *Die Künstlichen Edelsteine,* W. Diebner, Leipzig (1926)
34. H. Michel, Coloring precious stones by ray, *Jewelers' Circ.-Keyst.,* **95,** 10 (Sept 9, 1927)
35. F. H. Pough and T. H. Rogers, Experiments in X-ray irradiation of gem stones, *Amer. Min.,* **32,** 31 (1947)
36. F. H. Pough, The coloration of gemstones by electron bombardment, *Z. dt. Ges. Edelsteink.,* **20,** 71 (1957)
37. K. Nassau, The effect of gamma rays on the color of beryl, smoky quartz, amethyst and topaz, *Lap. J.,* **28,** 20 (1974)

Chapter 3

Heat Treatments

'Tis the voice of the lobster; I heard him declare,
'You have baked me too brown, I must sugar my hair'.
Lewis Carroll

The Beginnings of Heat Treatments

It can only be guessed when the enhancement of gemstones by heat treatment first evolved. Treatments were in full use in Greece and Rome well before the Christian Era, as discussed in Chapter 2. Heating above the boiling point of water does not appear to be mentioned specifically, although it is implied in the papyrus *P. Holm.*, where the loosening (cracking) of stones in preparation for dyeing is described, as discussed in Chapter 5. The use of shaped gemstones in jewelry had already reached a highly sophisticated level by about 3000 BC, from which time specimens have survived from Ur in Sumeria (Mesopotamia), from Harappa in India (now West Pakistan), and from the early dynasties of Egypt[1]. Red agate and carnelian that had been heated are reported to have been made in India by about 2000 BC and were found in Tutankhamen's tomb (about 1300 BC), as well as in other tombs[2].

In the numerous conquests and sackings of towns there were probably many occasions when slaves found the charred remains of precious belongings while poking through the burned remains of their regal masters (only rulers and priests owned or wore jewelry in earlier days). On some such occasions the discovery of a partially 'cooked' gem, which had actually improved in color, can be visualized. And there could have been an occasion when this information filtered back to one of the two preservers of technical information of those days, the priests and the artisans. Whichever received the information, it was without doubt the artisan who tried this process on some of the precious materials entrusted to him for fashioning into jewelry. Thus may have begun the history of heat treatments, assuming the artisan found some success before doing enough damage to forfeit his or her life.

Over the years there were a number of early technological studies on the effect on gems and minerals of heating in various atmospheres and at various temperatures, often with the aim of understanding the causes of the colors, as for example in reports by Simon[3] and Weigel[4]. Yet Wild of Idar complains[5] as late as 1932 that:

> Until very recently, the heating of gem stones was practiced in a crude and inefficient manner; the stones were generally placed into fine, dry sand and heated in a crucible over either the fire in a stove or in an old fashioned muffle furnace. No records of temperature were taken and no attempts were ever made to place a thermometer alongside the stones as it was a firm belief that the crucible must be entirely covered and not the slightest opening should exist to permit the 'entrance of air' into it. In talking with various old timers, the author found that it is the general belief that there is no air within the crucible because it is entirely filled with sand and tightly covered. The meaning of a vacuum is somewhat strange to these people who are not accustomed to thinking in physical terms.
>
> Altogether, the burning was a delicate enterprise and often resulted in failure by over or underheating if not partly or completely cracking the stone which was to be augmented in value. There were, of course, some men with long experience who took certain precautions which insured them a higher success than the beginner could ever hope to achieve.

Wild carried out careful experiments and reported the specific data given in *Table 3.1*. It is interesting to note that he makes no mention of the possibility of using a reducing environment (*see below*). Even today most heat treatments are still performed in primitive equipment with very uncertain results.

Table 3.1 Heat treatments reported in 1932 by Wild[5]

	Gems not turned by heat	
	Ruby, Australian sapphire, spinel, emerald, chrysoberyl, peridot, garnet, kunzite	
	Gems changed by heat	
	Color produced	*Temp. (°C)*
Blue Ceylon sapphire	Faint yellow–white	400
Purple ceylon sapphire	Pink	450
Precious topaz (orange)	Pink	500
Green beryl	Blue (aquamarine)	420
Yellow beryl	Light blue–white	400
Dark red tourmaline	Pink	550–600
Blue–green tourmaline	Clearer green	650
Smoky green tourmaline	Clearer green	600–650
Smoky quartz	White	275–300
Smoky yellow crystal	Yellow–orange	250–350
Certain amethysts	Orange–yellow	500–575
Salmon-colored beryls	Clear pink (morganite)	400
Brownish beryl	Pink	400
Zircon-green-blue	Clear blue	380–500

Since the 1970s there has been a tremendous increase in the use of heat treatments, particularly as applied to the corundum family although, in retrospect, it is clear that this technique has been steadily developed over the years. Thus, R. Crowningshield (unpublished observation) has recently noted the almost total absence for some time of the brownish or purplish Thai rubies, indicating that almost all such stones had been heat treated to remove the brownish or purplish shade, without any note being taken of this activity. The wide range of changes that is being induced today by heat treatments is illustrated in *Table 3.2*.

Table 3.2 The major changes on heating gemstones

Material	*Changes*
Amber	Darker; pressed (reconstructed)
Beryl, aquamarine	Green to blue; yellow to colorless; orange to pink
Corundum, blue sapphire	Darken blue; lighten blue; add or remove asterism; add color
Corundum, colorless	To yellow, green, or blue; add color
Corundum, ruby	Purple or brownish to red; pink to orange; add or remove asterism; add color
Diamond	Change color of irradiated stones
Ivory	Darker
Quartz, amethyst	To colorless, yellow, brown, green, or milky
Quartz, carnelian, tiger's eye, etc.	Yellow or brown to red–brown or red
Quartz, smoky	Paler, greenish yellow, or colorless
Spodumene and kunzite	Paler or colorless; purple, blue, or green to pink or violet
Topaz	Yellow, brown, or blue to colorless
Tourmaline	Blue or blue–green to green; red to paler or colorless
Zircon	Brown to reddish, colorless, or blue; green to blue or yellow
Zoisite	To deep purple–blue

Heat Treatments and Furnaces

There are a multitude of names both for the process of heating as well as for the apparatus in which it is performed. One can speak of annealing, baking, browning, combusting, cooking, firing, frying, heating, incinerating, pyrolizing, roasting, scorching, searing, soaking, tempering, toasting, and so on. These operations occur in a brazier, cooker, furnace, heater, kiln, lehr, muffle, oven, pot, retort, roaster, and the like, a list that also includes terms usually found only in crossword puzzles, such as calcar, oast, and salamander. Some of these may be combined with a variety of qualifiers, such as bath, bell, box, car-bottom, continuous, crucible, elevator, hearth, horizontal, muffle, pit, pot, reverberatory, soaking, split, tube, tunnel, or vertical. There are subtle differences involved in some of these terms, but they are unimportant in the present context; for simplicity, 'heating' in a 'furnace' is used herein. There are also many different heat-generating agents, including various gases, oil, charcoal, wood, coal, coke, a variety of applications of electricity, as well as optical-image systems that use the sun or other intense light sources.

Heat treatments can be as simple as placing a specimen of brown topaz into a glass test-tube and delicately holding it over the flame of a small alcohol burner. This has the advantage that the change in color can be seen and heating stopped as soon as the desired change occurs, but care must be used to apply the heat very gradually. Another simple technique is to place the specimen into the kitchen oven or bury it in the center of a charcoal brazier; the 'industrial' version of the latter is the charcoal-filled oil drum. A quaint technique mentioned in early works is to place a gemstone into the center of a dry sponge and set the sponge on fire, or to place it into a date or a fig and burn this in a fire! There are severe temperature limitations on such primitive techniques and more sophisticated arrangements may need to be used. Descriptions of various types of furnace materials and constructions, as well as of the different techniques used to measure and control the temperature, are given in Appendix A, where references are also given. *Figure 3.1* illustrates a bellows-operated furnace, as used in Agricola's time (1556)[6].

Figure 3.1 A bellows-operated furnace from Agricola's 1556 book on metallurgy

Most gemstone materials require heating at a rather gentle rate so as to avoid fractures. This is usually achieved by burying the material in some inert powder or placing it in a series of nested crucibles (usually of high-purity alumina for use up to 1900 °C, as in *Figure 3.2*), so that the heat penetrates slowly to and into the gemstone. The whole arrangement is then placed into the cold furnace which is not opened again until the heating is completed and the stone has cooled to room temperature. Heating past 100 °C should be done particularly slowly if moisture may be present, as in some of the chalcedony group (agate, carnelian, etc.). Various materials have been used for burying, including the iron filings and unburned lime of Porta, *see* Chapter 2, and the mixed iron filings and sand of Kluge, also mentioned there. Pure quartz sand is usually very satisfactory at lower temperatures (say up to 1100 °C or a little above) while alumina powder (powdered corundum, 'Linde A', 'boule powder', etc.) can be used up to almost 2000 °C. Charcoal, or graphite powder or granules, can be used at all temperatures where reducing conditions (*see below*) are required or acceptable.

Figure 3.2 Nested alumina crucibles used for heat-treating yellow sapphires. Photograph by courtesy of the Gemological Institute of America

One way in which gemstone materials may self-destruct during heating is by the enlargement of existing cracks; it may therefore be advisable in some instances to trim away all damaged regions. Since this trimming process is part of the gemstone-shaping process, heating is frequently performed on the preform or the finished gem; repolishing may then be required.

Another cause of fractures, either local or overall, is the presence of inclusions of various types. If an inclusion has a higher thermal expansion than the surroundings, a large stress will result if the material is heated above the original formation temperature, as illustrated in *Figures 3.3(a)* and *(b)*. Below the

(a) 

(b) 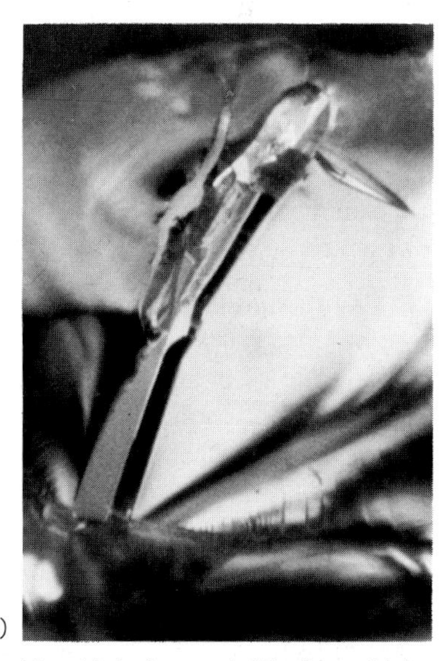

Figure 3.3 A needle-shaped inclusion in a sapphire (a) before and (b) after a high-temperature heat treatment. Photographs by courtesy of the Gemological Institute of America

formation temperature the inclusion will have shrunk into a smaller space. Chemical decompositions or reactions that yield gases or bulkier products can also produce damage. In some instances this type of damage can reveal that a heat treatment has been used, as discussed in Chapters 6 and 7.

Table 3.3 Temperature conversion

Degrees C	*Degrees F*	*Degrees C*	*Degrees F*	*Degrees C*	*Degrees F*
50	122	750	1382	1450	2642
100	212	800	1472	1500	2732
150	302	850	1562	1550	2822
200	392	900	1652	1600	2912
250	482	950	1742	1650	3002
300	572	1000	1832	1700	3092
350	662	1050	1922	1750	3182
400	752	1100	2012	1800	3272
450	842	1150	2102	1850	3362
500	932	1200	2192	1900	3452
550	1022	1250	2282	1950	3542
600	1112	1300	2372	2000	3632
650	1202	1350	2462	2050	3722
700	1292	1400	2552	2100	3812

Since all temperatures in this book are given in degrees celsius, °C (identical in value with the older 'degrees centigrade'), a conversion is given in *Table 3.3* for those readers more comfortable in degrees farenheit, °F. The melting point of corundum, 2050 °C, represents the highest temperature that might ever need to be used in the heat treatment of gemstones.

Heat-treatment Conditions

The important factors in specifying the conditions for the heat treatment of a gemstone material are the following:

(1) The maximum temperature reached.
(2) The time for which the maximum temperature is sustained.
(3) The rate of heating to temperature.
(4) The rate of cooling down from temperature and any holding stages while cooling.
(5) The chemical nature of the atmosphere.
(6) The pressure of the atmosphere.
(7) The nature of the material in contact with the gemstone.

These seven factors may even vary during the course of a specific heat treatment. In any given treatment several of these factors may not be important; the absence of any mention in the discussions of Chapter 7 implies that this is the case (or merely that the author is not aware of the importance of such factors!).

Exact conditions for heat treatments cannot usually be specified for two reasons: first, because most heat-treatment processes are held as closely guarded secrets and have never been fully revealed; and second, and most importantly, because of the wide variability of natural materials, as discussed in Chapter 1.

To give one specific example, the blue color of sapphire as it is found in nature is derived from a subtle interaction between two impurities, iron and titanium. This color can be further modified by the presence of other impurities, such as the red-causing chromium, or even by the white silk- and asterism-producing titanium itself; this last factor is controlled in part by the heating and cooling conditions to which the material was last exposed in its geological history. The exact shade of blue also depends not only on the relative amounts of iron and titanium present, but also on the valence states involved, namely ferrous, Fe^{2+}, and ferric, Fe^{3+}, as well as titanous Ti^{3+}, and titanic, Ti^{4+}, states; this is controlled by the oxidizing–reducing conditions during formation and subsequent heating and cooling in nature. The exact appearance of any specific, as-mined, Fe–Ti-coloured blue sapphire, which can range from almost colorless via yellow, green, and blue to almost black with red, purple, brown, or milky overtones, either clear or combined with silk or asterism, is not indicative of an exact composition, but could be derived from a broad range of different compositions and past environmental histories. In attempting to produce a specific color enhancement in such a gemstone by a heat treatment, it is obvious that a wide range of conditions might have to be tried to find the correct process, which could be quite different for a similarly appearing stone from a different locality containing different impurity concentrations or having had a different history.

In some instances, no enhancement may appear possible at first glance, as in a ruby, where the color is caused by chromium and where a heat treatment cannot change the valence state or the color of the Al_2O_3–Cr^{3+} combination. Yet here, too, the color may contain a brown, purple, or milky component derived from iron and titanium impurities which could be enhanced by a heat treatment, as illustrated in *Plate XIV*, with the same complexities present as discussed above.

The Effects of Heat

Heat can have many different effects; the nine that are most important in gemstone materials are summarized in *Table 3.4*. The examples given in this table and in this chapter are only illustrative; full details are in the listing of Chapter 7. Omitted from *Table 3.4* are those effects of heat that are completely reversible, one example being the turning green of a ruby when heated to red heat, where the original color returns when the ruby is cooled back to room temperature.

The first item of *Table 3.4* is the darkening effect, which is sometimes used to 'age' amber or ivory. This is the equivalent of a gentle charring, at least in part. However, since amber has been reported to darken even on storage in the dark[7], it is likely that air oxidation of the organic compounds present could also be involved; such an oxidation would, of course, be speeded up by a gentle heating.

The second item of *Table 3.4*, the destruction of color centers by heating, is described in detail in Chapter 4. This may result in bleaching or fading, as when brown topaz, blue topaz, irradiated yellow sapphire, red tourmaline, or smoky quartz are heated. In other instances there may be a color change, as when some amethyst turns to a very pale citrine (but also *see below*) or to 'greened amethyst', or when the bleaching of a brown topaz reveals the presence of a previously hidden chromium-derived color in a 'pinked' topaz. These color changes can usually be reversed by an irradiation treatment, as described in Chapter 4; more details are given for specific materials in Chapter 7.

Table 3.4 The effects of heat on gemstone materials

Effect	*Mechanism*	*Examples*
(1) Darkening	Gentle charring and/or oxidation	'Aged' amber and ivory
(2) Color change	Destruction of color center	Blue or brown topaz and zircon to colorless; 'pinked' topaz; amethyst to pale yellow or green; smoky quartz to greenish yellow or colorless
(3) Color change	Change in hydration, aggregation	Carnelian to orange, red, or brown; sapphire to deep yellow or orange
(4) Structural change	Reverse the irradiation-induced metamict state	'Low' zircon to 'high' zircon
(5) Color change	Oxidation state change, usually with oxygen diffusion	Green aquamarine to blue; amethyst to deep citrine; colorless/ yellow/green/blue sapphire; brown or purple ruby to red
(6) Structural change	Precipitation or solution of a second phase	Development or removal of silk or asterism in corundum
(7) Color addition	Impurity diffusion	Diffused color and asterism in sapphire
(8) Cracking	Rapid change of temperature	Fingerprints in sapphire; 'crackled' quartz
(9) 'Reconstruction' and clarification	Flow under heat and pressure	Reconstructed and clarified amber, reconstructed tortoise shell

The color changes derived from a change in hydration or aggregation, the third item of *Table 3.4*, usually involve iron impurities. If limonite (yellow, hydrated iron oxide), $Fe_2O_3.xH_2O$, the pigment 'ochre' of the artist, is heated, the deep orange, brown, or red color of rust or of 'burned ochre' is obtained, which is the same material as the mineral hematite. A somewhat similar change occurs when some gray-to-yellow-to-brown iron-containing quartzes, such as agate, carnelian, or tiger-eye, are heated to produce deep brown-to-red colors. If the slightly heated, iron-containing pale citrine described in the previous paragraph is heated more strongly, or if natural, very pale yellow or green, iron-containing sapphire is similarly heated, there is a change in the individual iron atoms distributed throughout the crystal, which give the yellow or green colors. These atoms may move together to form molecules and then small aggregates of iron oxide, thus leading to the deep yellow, orange, or reddish brown colors of iron oxide. These steps are not reversible, as is the bleaching of color centers.

The reversal of the metamict state, item 4 of *Table 3.4*, is discussed under zircon in Chapter 7. The last five items of this table deserve the following more expanded discussions.

Oxidizing–Reducing Conditions and Gas Diffusion

When any oxygen-containing substance, say a crystal of corundum (aluminum oxide, the dominant component of ruby and sapphire), is heated in an enclosed space to a very high temperature, a little oxygen gas is either emitted or absorbed. If enough of the oxide is present and if this experiment is continued long enough, the composition of the atmosphere in contact with the oxide stabilizes at a certain oxygen concentration, called the 'equilibrium partial oxygen vapor pressure', often written p_o. If the atmosphere is now changed by adding more oxygen at the same temperature, then the circumstances are designated 'oxidizing' and extra oxygen enters the crystal. This may correspond to a movement of only a few atoms of oxygen, which may not even produce any measurable change; nevertheless, it happens to some extent. Alternatively, if the oxygen is reduced below p_o, then at least some oxygen leaves the crystal and we speak of 'reducing' conditions.

As the temperature is raised, p_o increases; accordingly, with a constant oxygen pressure P, it is possible that there will be a change from oxidizing, where p_o is less than P, to reducing conditions, where p_o is larger than P, just by raising the temperature.

For most oxide gem-materials, such as Al_2O_3, at temperatures below the melting point, air is on the oxygen-rich side and therefore oxidizing, but only slightly so. Pure oxygen is normally the most oxidizing atmosphere one can use, except that it would be possible to go somewhat further by increasing the pressure and thereby the density of the oxygen.

A non-reactive gas, such as nitrogen, is reducing, but only slightly so. Any gas that reacts with oxygen and so removes it is, of course, strongly reducing; one example is hydrogen gas, which forms water:

$$2H_2 + O_2 \rightarrow 2H_2O \qquad (3.1)$$

Hydrogen by itself is rather hazardous since even a small leak can produce a violent explosion. Very useful is 'forming gas', a much less explosive mixture of hydrogen and nitrogen, which is not even inflammable up to 8 percent hydrogen concentrations and is still strongly reducing. A frequently used reducing atmosphere is obtained with charcoal or other carbon-containing substances, such as sugar, oil, or glycerin, which produce the equivalent of carbon-containing charcoal when heated. In a confined space made, for example, by placing the material into a crucible with an excess of carbon and sealing it with some type of cement, the residual oxygen present reacts to give carbon monoxide:

$$2C + O_2 \rightarrow 2CO \qquad (3.2)$$

which can then remove any freshly liberated oxygen by reacting to form carbon dioxide:

$$2CO + O_2 \rightarrow 2CO_2 \quad (3.3)$$

This carbon dioxide can now react with more carbon to re-form the carbon monoxide:

$$CO_2 + C \rightarrow 2CO \quad (3.4)$$

One way in which reducing conditions can be achieved is to have the material exposed to the direct flame when gas, propane, gasoline, etc., are being used, as described in Appendix A. Restricting the amount of air or oxygen used in the flame gives a sooty flame, which then supplies the carbon required for reactions (3.2) and (3.4).

To affect the whole of a specimen, oxygen must diffuse all the way in to the center of the specimen (or out of it), a distance of perhaps a centimeter. This takes time, possibly a few hours at temperatures well over 1000 °C. Such rapid diffusion can occur since the oxygen does not need to move the whole distance to produce the desired effect because of the nature of oxide structures. Consider a schematic representation of the aluminum oxide crystal-structure, in which there is an oxygen atom missing at the left, *Figure 3.4*, with an oxygen molecule at the right in the atmosphere just above the surface. By a co-ordinated series of jumps of all the

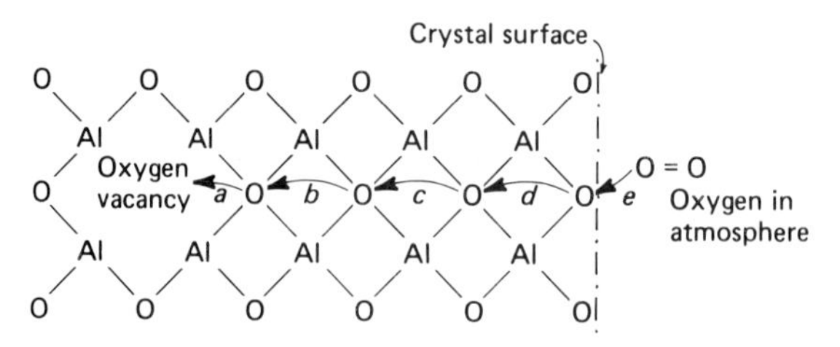

Figure 3.4 The diffusion of oxygen from the atmosphere to fill an oxygen vacancy in corundum, aluminum oxide

intermediate oxygens, as shown by the arrows, the vacancy can be filled, with each oxygen moving no more than a few nanometers (one-millionth of a millimeter). When using hydrogen, diffusion can be even more rapid, since the hydrogen atom is so much smaller than the oxygen atom and can therefore pass quite rapidly between the other atoms. When the diffusion of added impurities is involved, as in the next section, then the motion required is much larger and diffusion rates are very much slower.

The presence of varying amounts of oxygen can affect the valence state of an impurity, such as iron present in Al_2O_3. Iron usually exists either in the ferrous state, Fe^{2+}, i.e. as FeO, or in the ferric state as Fe^{3+}, i.e. as Fe_2O_3. At the high temperature of crystal growth the iron is normally in the divalent state as Fe^{2+} or FeO. When such an Fe^{2+} replaces one of the Al^{3+} in corundum, there is one oxygen vacancy for every two Fe^{2+}, to maintain an electrically neutral crystal. In corundum this may yield an almost colorless crystal or perhaps a very pale green at higher concentrations. The composition can be written as:

$$(1-x)Al_2O_3 + 2xFeO \rightarrow Al_{2-2x}Fe_{2x}O_{3-x} \quad (3.5)$$

Consider now the oxidation of Fe^{2+} to Fe^{3+} by gaseous oxygen, that is:

$$4Fe^{2+} + O_2 \rightarrow 4Fe^{3+} + 2O^{2-} \quad (3.6)$$

or, as the oxide:

$$4FeO + O_2 \rightarrow 2Fe_2O_3 \quad (3.7)$$

or, in corundum:

$$2Al_{2-2x}Fe_{2x}O_{3-x} + xO_2 \rightarrow 2Al_{2-2x}Fe_{2x}O_3 \quad (3.8)$$

All three equations are equivalent and result in an electrically neutral state containing only Fe^{3+}; in corundum the result now corresponds to Fe_2O_3 present in Al_2O_3, having a pale-to-medium yellow color, depending on the concentration. There is now no need for oxygen vacancies.

By heating under reducing conditions, say in hydrogen or with carbon monoxide, the reverse process can be produced:

$$Fe_2O_3 + H_2 \rightarrow 2FeO + H_2O \quad (3.9)$$

or:

$$Al_{2-2x}Fe_{2x}O_3 + xH_2 \rightarrow Al_{2-2x}Fe_{2x}O_{3-x} + xH_2O \quad (3.10)$$

and:

$$Fe_2O_3 + CO \rightarrow 2FeO + CO_2 \quad (3.11)$$

or:

$$Al_{2-2x}Fe_{2x}O_3 + xCO \rightarrow Al_{2-2x}Fe_{2x}O_{3-x} + xCO_2 \quad (3.12)$$

If heated more strongly, as discussed in connection with process 3 of *Table 3.4*, the Fe_2O_3 could aggregate to form molecules and then particles of hematite, thus producing a yet deeper yellow-to-brown color.

Oxidation–Reduction in Blue Sapphire

With the preceding section as background, the combination of iron plus titanium, which gives sapphire its blue color, can now be explored. Consider these two impurities, present at a concentration of a few hundredths of one percent in the fully oxidized state. Iron is present as Fe^{3+} substituting for Al^{3+}, giving at most a trace of a pale yellow color. Titanium is present as Ti^{4+} substituting for Al^{3+} with an arrangement, discussed below, which maintains electrical neutrality. By itself, Ti^{4+} produces no color.

Now consider what happens when a slight reduction is produced, say by a heat treatment using hydrogen or charcoal. This converts some of the Fe^{3+} into Fe^{2+}, following equations (3.9) to (3.12). With both Fe^{2+} and Ti^{4+} now present, a new interaction becomes possible; this is called charge transfer and involves the hopping of one electron from the Fe^{2+} to the Ti^{4+}:

$$Fe^{2+} \rightarrow Fe^{3+} + e^- \quad (3.13)$$

and:

$$Ti^{4+} + e^- \rightarrow Ti^{3+} \quad (3.14)$$

or the combination of equations (3.13) and (3.14):

$$Fe^{2+} + Ti^{4+} \rightarrow Fe^{3+} + Ti^{3+} \quad (3.15)$$

The products on the right of equation (3.15) are in a higher energy state than those on the left, as indicated in *Figure 3.5*, and the reaction only occurs when light falls on the specimen and is absorbed to supply the energy required by the process. Only some of the light is absorbed; the light remaining gives the blue color to the sapphire. The resulting absorption spectrum is shown in *Figure 3.6*, where the absorption band labeled 'b' produces the blue color.

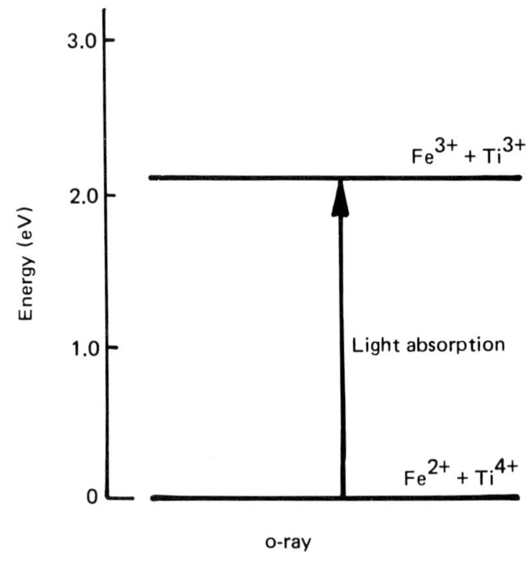

Figure 3.5 Energy levels involved in the charge-transfer mechanism that gives the blue color in sapphire. From K. Nassau, *The Physics and Chemistry of Color*, Wiley, New York (1983)

Shortly after absorbing the light, the reverse reaction of equation (3.15) occurs, and the energy absorbed is emitted as heat which warms the sapphire a little. There are other features in *Figure 3.6* which derive from iron by itself, from iron–iron charge transfer, and so on. Although only one is shown, there are actually two curves to the absorption spectrum of *Figure 3.6*, corresponding to different polarizations of the light absorbed and leading to the dichroism of sapphire. If reduction is continued, the color deepens as more Fe^{3+} is converted into Fe^{2+}.

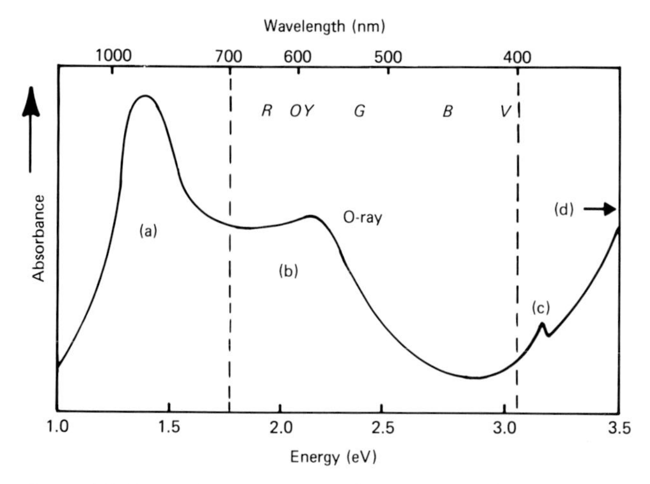

Figure 3.6 The ordinary-ray absorption spectrum of blue corundum showing absorptions derived from: (a) Fe^{2+}–Fe^{3+} charge transfer; (b) Fe^{2+}–Ti^{4+} charge transfer; (c) ligand-field effect in Fe^{3+}; and (d) O^{2-}–Fe^{3+} charge transfer. After G. Lehmann and H. Harder, *Amer. Min.*, **55,** 98 (1970)

Note that there is a tremendous intensification in color in going from the very pale yellow of Fe^{3+} to the deep blue of the Fe^{2+}–Ti^{4+} combination even though only a part of the iron present partakes, so that the effective concentration is much smaller. The Fe^{3+} color is produced by a ligand-field (crystal-field) effect in which the transitions are 'forbidden' by certain 'selection rules', and therefore lead to very weak absorptions. Deep colors result only when several percent of such an impurity is present as, for example, in ruby. Charge-transfer absorptions are 'allowed', however, and are therefore 100 to 1000 tines as intense; this leads to saturated colors even at the one-hundredth percent impurity level, as in blue sapphire.

Oxidation reverses these changes and can lighten the blue color or even remove it so that a very pale yellow or colorless material may result. Note that the production of blue and white two-color sapphires by Porta in 1658, given in Chapter 2, pp. 17–18, is exactly this process with the lack of access of air to the coated half of the stone preventing oxidation and loss of the blue color.

If the iron concentration is high relative to the titanium, it is then possible to obtain by reduction a yellow, a bluish green, or even a green color derived predominantly from Fe^{2+}. If the titanium concentration is relatively high, then the possibility of asterism arises, as discussed below.

The first person to establish the presence of both iron and titanium in natural blue sapphire was Professor A. V. L. Verneuil, the discoverer of the Verneuil flame-fusion process for the synthesis of ruby and sapphire. He found that he was unable to obtain the desired blue color without the presence of both the elements iron and titanium[8–10]. It took many years before the significance of the Fe–Ti charge-transfer was appreciated[11], and even today not all aspects of the sapphire absorption spectrum are fully understood. The reader in search of the full complexity can find summaries and some of the many relevant references in papers by Schmetzer and Bank[12, 13], although it is obvious that not all of the varied possibilities suggested by the many workers who have explored different parts of this veritable jungle can be correct, since there are incompatibilities among the various interpretations.

It has also been reported[12] that there can be some increase in the concentration of iron near the surface during heat treatment at temperatures near 1000 °C, particularly when reducing conditions are present. Presumably, diffusion moves the iron outward from the interior of the stone, a process which is said to be intensified by the presence of some sodium carbonate on the surface; this latter material is often added during heating, mainly to prevent cracking, as discussed in Chapter 7. Since the surface is repolished after heat treatment, this process is supposed to produce a lightening of the color[12].

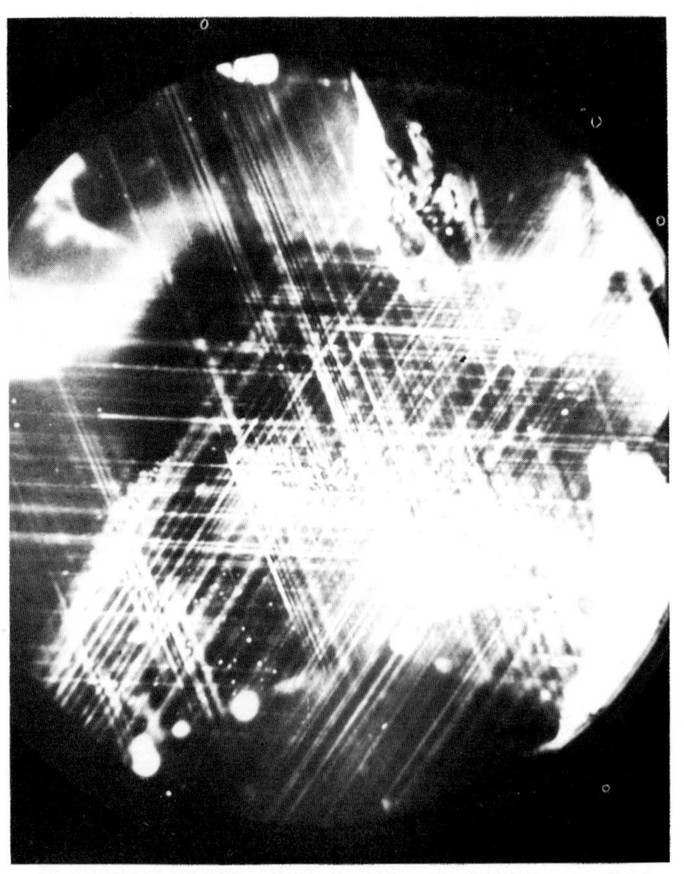

Figure 3.7 Rutile needles in a star sapphire (magnification 1000x). Photograph by courtesy of the Gemological Institute of America

Silk and Asterism in Corundum

Many impurities can produce a milkiness or silkiness in crystals, but the dominant cause in corundum is titanium, which also produces the asterism shown in *Figure 3.7*. The evidence that asterism was caused by needles of rutile, TiO_2, was summarized by Tait[14]. One of the most direct determinations was that of Frondel[15], who ground-up star corundum and removed the corundum with a heavy liquid; the author also briefly studied a natural star corundum[16]. More recently Phillips *et al.*[17] presented a detailed examination with many references, although they erroneously misrepresented at least one of the earlier reports[16].

It was the successful duplication of natural, asteriated corundum by Burdick and Glenn in a patent assigned to the Linde Co[18] that led to a detailed understanding of the mechanism involved in the formation of asterism. It was found that when any type of corundum is grown by the Verneuil technique with about 0.2 percent rutile, TiO_2, added to the Al_2O_3, then the resulting crystal boule was clear and showed no evidence of the presence of the titanium if it was cooled fairly rapidly in the normal way. It should be noted that this is a much higher concentration than that needed to produce the blue color when combined with iron.

If the boule is now re-heated to 1100–1500 °C (or if cooling from growth is interrupted at this temperature), then over a period of time tiny needles of rutile form within the corundum. At the lower temperatures several days are required, but only hours at the higher temperatures. When the boule is cooled to room temperature and shaped in the form of a cabochon, the asterism is seen as the usual six-rayed star. There appears to be a twin plane within almost all the needles[17], and the orientation constraint leading to the three parallel sets of needles at 120 degrees within the basal plane from which the six rays are reflected is dependent on complex relationships between the crystal structures of rutile and corundum.

At temperatures above about 1600 °C, the titanium present at, say, 0.2 percent concentration is completely soluble in Al_2O_3 in the form of what is usually called a 'solid solution'. The titanium is present mostly as Ti^{3+} substituting for Al^{3+}, i.e. as Ti_2O_3 in Al_2O_3. What actually happens is that the rutile, TiO_2, which is added to the feed powder, loses oxygen at the very high melting temperature:

$$4TiO_2 \rightarrow 2Ti_2O_3 + O_2 \qquad (3.16)$$

A little of the titanium is also present as Ti^{4+} or TiO_2, but some mechanism must maintain charge neutrality. There is one Al vacancy for every three Ti^{4+}, so that these three Ti^{4+} take the place of four Al^{3+}; note that either set has a charge of 12+. Also note that if this is viewed as $6TiO_2$ replacing $4Al_2O_3$ with 6Ti replacing 8Al, then either set contains 12 oxygens. There is no change in structure when this material is cooled rapidly to room temperature. This process can be written as:

$$(1-2x)Al_2O_3 + 3xTiO_2 \rightarrow Al_{2-4x}Ti_{3x}O_3 \qquad (3.17)$$

It should, however, be noted that an alternative mechanism, involving interstitial oxygen, has also been proposed[17]:

$$(1-x)Al_2O_3 + 2xTiO_2 \rightarrow Al_{2-2x}Ti_{2x}O_{3+x} \qquad (3.18)$$

The final word on this possibility is not yet in.

Phase-diagram research has shown that at temperatures below about 1600 °C there is a limited solubility of titanium oxide in aluminum oxide[19]; accordingly, if the sample is now held at, say, 1300 °C for some time, two processes occur. First, oxygen diffuses in and converts the Ti^{3+}, as in Ti_2O_3, to Ti^{4+}, as in TiO_2:

$$2Ti_2O_3 + O_2 \rightarrow 4TiO_2 \quad (3.19)$$

The quantity of TiO_2 now present is too large to remain in solution and the excess precipitates out of solid solution in the form of TiO_2 needles having the rutile structure, following the reverse reaction of equations (3.17) or (3.18). Not all the TiO_2 precipitates, however, and several hundredths of one percent remain in solution as Ti^{4+}, so that they can interact with Fe^{2+}, if present, to give the blue color.

The rutile needles may range in size from a fraction of a micrometer in diameter to being visible to the naked eye; the best optical effects are seen at the lower end of this range. Even for small needles, it is obvious that there must be much atomic motion to permit the Ti atoms to diffuse toward, and the Al atoms to diffuse away from, the region where a needle is forming; only smaller motions are needed from the oxygen atoms, which can also move more easily by the co-operative phenomenon described above and shown in *Figure 3.4.* This is the reason why both the asterism-forming as well as the asterism-removing steps take some time, even at very high temperatures. The time is not excessive, however, because the distances involved, those between the needles, are small, typically one micrometer or so. This type of impurity diffusion is even more important when color or asterism is diffused into the surface of a stone, as discussed below.

To remove asterism or silk caused by TiO_2 in corundum, it is only necessary to reheat the specimen above about 1600 °C and wait sufficiently long for the TiO_2 to diffuse away from the needles, lose oxygen by reaction (3.16), and dissolve in Al_2O_3 mostly as Ti_2O_3, as described above. The material is then cooled relatively rapidly so that the time for oxygen in-diffusion and needle growth below about 1600 °C is not available. In this way, silky, milky, hazy, or whitish rubies and sapphires in which the imperfections are derived from titanium can be made clear, asterism can be removed, or asterism developed if the required Ti-content is present and nature omitted to provide the needle-growing conditions. It is even possible to take corundum that has needles which are too coarse to give a good star reflection, dissolve the needles at a high temperature, and then re-form them at the desired size at a lower temperature, so as to obtain an improved star. If the imperfections are not derived from titanium but from other impurities, an improvement, but probably not as complete, is still possible from a heating step, since almost all substances dissolve more completely at higher temperatures and take time to precipitate again at lower temperatures.

Impurity Diffusion

When the attempt is made to diffuse color- or asterism-producing impurities into a corundum, then very much longer diffusion times are required. In this process (7 in *Table 3.4*), say for diffusing-in chromium to produce a red color, the impurity atom must replace an Al at the surface, with the Al moving outward, and then there follows a series of Al–Cr interchanges in random directions; with time the net result is that the Cr atoms move inward, but only very slowly. As might be expected, much higher temperatures and longer periods are required for this process, with its need for long-distance motion, than with the previously discussed diffusions.

This type of diffusion process in corundum was developed by Eversole and Burdick[18, 20] and Carr and Nisevich[21–23] at the Linde Co, with the aim of making or improving synthetic stars and later of adding color as well. Their patents describe the processes in detail. For stars, the preformed or finished gemstones are buried in a powder which is made by heating 35 parts of TiO_2 and 65 parts of Al_2O_3 powders for 4 h at 1300 °C, followed by grinding. Diffusion at 1750–1900 °C, typically at 1800 °C for 24 h, produces penetration of the titanium to a depth of about 0.004 to 0.01 inches (0.01 to 0.25 mm). Maintaining a reducing atmosphere during heating preserves the blue color of sapphire, if that is being used (and may even help the diffusion by creating Ti_2O_3 and oxygen vacancies, both of which may speed up the atomic motions).

Instead of merely burying the corundum in the powder, a slurry of the powder made up with water can be painted on the surface of the stone, which is then left to dry. Other impurities, usually at the several percent level, are added to Al_2O_3 to produce the powder used for the addition of colors. Adding chromium gives a ruby red, adding iron plus titanium (or either one if the other is already present in the stone) gives a blue, iron plus chromium an orange to pink (padparadscha), and so on. Since the penetrations of the asterism or the color are so shallow, great care must be taken in the subsequent polishing to ensure that the surface layer is not removed completely and all the color again lost!

One last use for diffusion is to even out the banding in Verneuil-grown corundum. This banding is derived mainly from impurity concentration fluctuations which are inherent in the Verneuil process[24]. Diffusion reduces the concentration variations with heating at a very high temperature for a very long time; the banding becomes much less prominent[25].

Heat-induced Cracking

Although cracking during heating is usually avoided, there are some occasions when this effect is actually considered desirable. One example is the 'crackling' of crystalline quartz by heating and dropping into water. The resulting cracks are considered attractive in themselves, resembling the naturally occurring iris quartz, with iridescent colors produced by the interference of the light waves after reflection from the walls of the thin cracks. By using a colored solution in the place of water, the resulting colored stone may be used as an imitation of other stones, such as ruby or emerald. The quaint quotation from Boyle's 1672 book, quoted in Chapter 2, gives a perfect description of this process, as do recipes in the much earlier *P. Holm.*, also given in Chapter 2.

More recently a somewhat similar process has been used on corundum, but for a different purpose; namely, to introduce 'fingerprint' inclusions[25, 26]. According to J. Coivula (personal communication) this process may have been performed in the following manner. A ruby, for example, may have fine cracks induced in it by an appropriately uneven heating, e.g. with an oxygen–hydrogen or an oxygen–acetylene torch; additional heating may then produce a partial healing of these cracks so that the final appearance resembles that of typical 'fingerprint' inclusions, as observed in natural corundum. The appearance of such fingerprints in a banded, synthetic Verneuil stone[25, 26] can be most confusing!

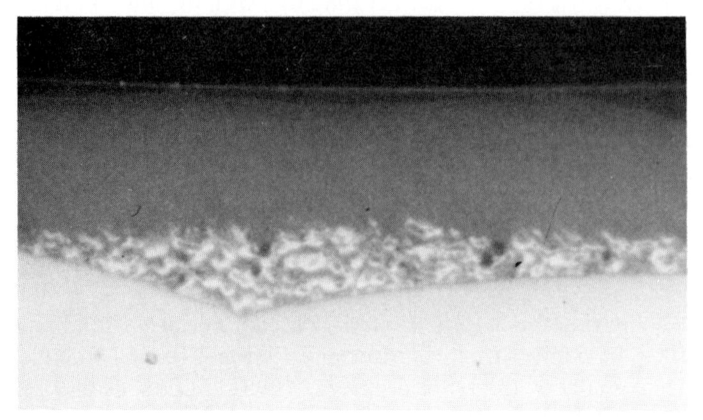

Figure 3.8 Double girdle on a sapphire showing damage on the surface; a part of the girdle was missed during repolishing after a high-temperature heat treatment. Photograph by courtesy of the Gemological Institute of America

Recent surveys of the technical aspects of the heat treatments used on corundum[27] and their detection[25] have been published. Cracking at inclusions is one way in which heat treatment can be identified; this has been discussed above in connection with *Figures 3.3(a)* and *(b)*. Another change often produced by heating to very high temperatures, particularly if in contact with a reactive chemical substance, is surface etching. If any surface is not repolished it can give a characteristic appearance, as shown in *Figure 3.8*.

The Reconstruction and Clarification of Amber

The term 'reconstruction' is often misused in the field of gemstones. Thus, it has been claimed that ruby was reconstructed just before the turn of this century when, in fact, synthetic rubies were being manufactured by an early precursor of the Verneuil technique. The falsity of these claims was exposed almost immediately, but had to be rediscovered later[28]. Also discussed in reference 28 is a much later report which described how sapphire was ground-up and fed through a Verneuil torch to once again grow sapphire; this, too, was called a reconstruction. Similar claims have been made for reconstructed turquoise made from the ground-up natural product, and so on.

The usual attitude of gemologists is that a truly reconstructed gemstone is one in which at least a part of the original material retains its properties, with some of its characteristic defects and inclusions remaining intact. On this basis it would appear that only pressed amber and the related tortoise-shell product qualify for the designation 'reconstructed'.

As discussed in Chapter 7 under amber, this material consists of a mixture of organic materials that softens above 150 °C and melts above 250 °C; these are the gradual changes that occur with any complex mixture. When pieces of relatively clear amber are heated to about 180 °C and simultaneously exposed to pressures reported to be in the 5000 to 120 000 pounds per square inch range (the higher pressure seems totally unreasonable!), the pieces fuse together to form a solid block. To prevent obvious discontinuities from revealing the boundaries between the original pieces, the pressure can also be used to force the pieces through a perforated steel plate having narrow slits; the product is not quite as clear but much more uniform. Air must be excluded to prevent deterioration. The names 'pressed amber' or 'ambroid' have also been applied to such truly 'reconstructed amber'. More details are given in Chapter 7. A small amount of oil, such as linseed oil, may be added when performing this process.

Much natural amber contains many small bubbles which make it opaque. A heating clarification process using rapeseed oil has been used since the seventeenth century to remove these bubbles. Details are given in two long papers by Dahms[29], which are essential for anyone interested; all later accounts are derived from these.

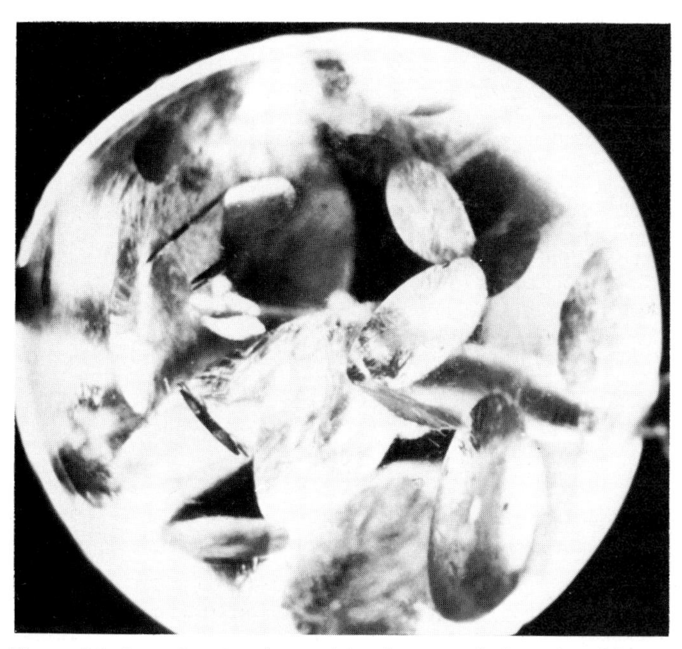

Figure 3.9 An amber bead containing 'sun-spangles' produced by a special heat treatment. Photograph by courtesy of the Gemological Institute of America

The basic process uses cold rapeseed oil, into which the amber is placed, and the mixture is heated very gradually over many hours. The mixture is kept at the simmering point for the time required, which depends on the type of amber and the size of the individual pieces, and then cooling is performed at the same slow rate as the heating. Some material may have to be treated several times for total removal of the haze.

It is often stated that the oil penetrates to fill the bubbles, but this seems most improbable from a technical point of view. It is much more likely that the elevated temperature, at which the amber is somewhat plastic, permits diffusion to the outside of the air and/or water that fill the bubbles. This view is confirmed by the fact that dry heating, if also done very slowly, has the same clarification effect as does heating in rapeseed oil[29]. With this aspect in mind, it seems probable that Pliny's boiling of amber 'in the fat of a suckling pig'[30] would also remove the haze and is a straight-forward description of a clarification process that is almost 2000 years old.

If the clarification process is not done quite correctly, then cracks can appear in the amber. This is sometimes done on purpose to produce 'stress figures' or 'sun-spangles', as shown in *Figure 3.9* and discussed in Chapter 7 under amber.

References

1. G. Gregorietti, *Jewelry Through the Ages,* Crescent Books, New York (1969)
2. H. Harder, Qualitätsbesserüng von Edelsteinen, insbesondere von Korunden, durch Wärmebehandlung ('Brennen'), *Aufschluss (Z. Freunde Min. Geol),* **33,** 213 (1982)
3. K. Simon, Beiträge zur Kentniss der Mineralfarben, *Neues Jb. Miner. Geol. und Paläont.,* **26,** 249 (1908)
4. O. Weigel, Über die Farbenänderung von Korund und Spinell mit der Temperatur, *Neues Jb. Miner.,* **48,** 274 (1923)
5. G. O. Wild, The treatment of gem stones by heat, *Rocks and Miner.,* **7,** 9 (1932)
6. G. Agricola, *De Re Metallica,* Basel (1556)
7. M. Bauer, *Precious Stones,* p. 539, Dover Publications, New York, 2 Vols (1968)
8. A. Verneuil, Observation sur une Note de M. L. Paris, sur la reproduction de la coloration bleu du Sapphir oriental, *C.R. Acad. Sci., Paris,* **147,** 1059 (1907)
9. A. Verneuil, Sur la reproduction synthétique du sapphir par la méthode de fusion, *C.R. Acad. Sci., Paris,* **150,** 185 (1909)
10. A. Verneuil, Sur la nature des oxydes qui colorent le sapphir oriental, *C.R. Acad. Sci., Paris,* **151,** 1063 (1910)
11. M. G. Townsend, Visible charge transfer band in blue sapphire, *Solid State Comm.,* **6,** 81 (1968)
12. K. Schmetzer and H. Bank, Explanation of the absorption spectra of natural and synthetic Fe- and Ti-containing corundums, *N. Jb. Miner. Abh.,* **139,** 216 (1980)
13. K. Schmetzer and H. Bank, The colour of natural corundum, *N. Jb. Miner. Mh.,* **1981 H.2,** 59 (1981)
14. A. S. Tait, Asterism in corundum, *J. Gemm.,* **5,** 65 (1955)
15. C. Frondel, Commercial synthesis of star sapphires and star rubies, *Trans. Amer. Inst. min. metall. Engrs,* **199,** 78 (1954)
16. K. Nassau, The cause of asterism in star corundum, *Amer. Min.,* **53,** 300 (1968)
17. D. S. Phillips, A. H. Heurer, and T. E. Mitchell, Precipitation in star sapphire, Parts I, II, and III, *Phil. Mag.,* **A42,** 385 (1980)
18. J. N. Burdick and J. W. Glenn, Jr, *Synthetic Star Rubies and Star Sapphires and Process for Producing Same,* US Patent 2 488 507, Nov 15 (1949)
19. R. J. Bratton, Precipitation and hardening behavior of czochralski star sapphire, *J. Appl. Phys.,* **42,** 211 (1971)
20. W. G. Eversole and J. N. Burdick, *Producing Asteriated Corundum in Crystals,* US Patent 2 690 630, Oct 5 (1954)
21. R. R. Carr and S. D. Nisevich, *Altering the Appearance of Sapphire Crystals,* US Patent 3 897 529, Jul 29 (1975)
22. R. R. Carr and S. D. Nisevich, *Altering the Appearance of Corundum Crystals,* US Patent 3 950 596, Apr 13 (1976)
23. R. R. Carr and S. D. Nisevich, *Altering the Appearance of Corundum Crystals,* US Patent 4 039 726, Aug 2 (1977)

24. K. Nassau, *Gems Made by Man,* Chilton, Radnor, PA (1980)
25. R. Crowningshield and K. Nassau, The heat and diffusion treatment of natural and synthetic sapphires, *J. Gemm.*, **17,** 528 (1981)
26. R. Crowningshield, Developments and highlights at GIA's lab in New York, *Gems Gemol.*, **16,** 315 (1980)
27. K. Nassau, Heat treating ruby and sapphire: technical aspects, *Gems Gemol.*, **17,** 121 (1981)
28. K. Nassau and R. Crowningshield, The synthesis of ruby, Part 2, The mystery of 'reconstructed' ruby solved, *Lap. J.*, **23,** 313 (1969)
29. P. Dahms, Mineralogische Untersuchungen über Bernstein, *Danzig, Schr.*, **8,** 97 (1892–1894), **9,** 1 (1895–1898)
30. D. E. Eichholz, *Pliny: Natural History,* Vol 10, Book 37, Chapter 11, p. 199, Harvard University Press, Cambridge, MA (1962)

Chapter 4

Irradiation Treatments

In the exceeding lustre and the pure
Intense irradiation . . . .
. . . . internal lightning
Percy Bysshe Shelley

Irradiation

The process of irradiation involves the exposure of a specimen to one of a variety of radiations. These include *rays*, constituting a large part of the electromagnetic spectrum, as well as various energetic *particles*, some of which may also be called 'rays'. Each of these sources has advantages and drawbacks. A summary is given in *Table 4.1*.

Table 4.1 Rays and particles used for the irradiation of gemstones

Radiation type	*Typical energy* (eV)	*Coloration uniformity*	*Power requirement*	*Induced radioactivity*	*Localized heating*
A. The electromagnetic spectrum					
Light	2–3	Variable	Low	No	No
Ultraviolet, s.w.	5	Variable	Low	No	No
X-rays	10 000	Poor	Medium	No	No
Gamma rays	1 000 000	Good	None	No	No
B. Particles					
Neutral: neutrons	1 000 000	Good	Very high	Yes	No
Negative: electrons	1 000 000	Poor	High	No	Very strong
Negative: electrons	10 000 000	Poor	High	Yes	Very strong
Positive: protons, deuterons, alpha particles, etc.	1 000 000	Poor	High	Yes	Some

In Appendix B is given a more detailed discussion of the production and detection of irradiation, including units of measurement and other relevant topics; this appendix is intended to complement the information presented in this chapter and gives additional references.

The Electromagnetic Spectrum

The full extent of the electromagnetic spectrum is shown in *Figure 4.1*. In the visible spectrum in the center there occur the electromagnetic vibrations called light; by these the colors of gemstones are perceived in a way described in Appendix C. Included in *Figure 4.1* are three of the many ways of specifying a specific light of a certain color: the frequency in hertz, abbreviated Hz (previously cycles per second); the wavelength in nanometers, abbreviated nm (1 nm = 10^{-7} cm = 10^{-9} m); and the energy in electron volts, abbreviated eV. Conversions among these various units are given in Appendix C, *Table C.2*.

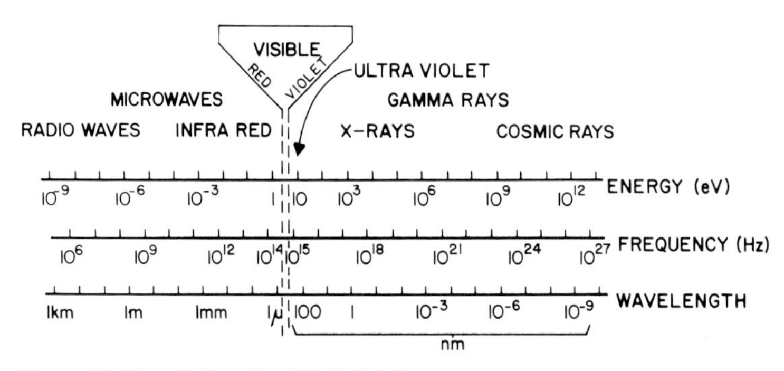

Figure 4.1 The electromagnetic spectrum

Some colors may be altered by visible light of energy 1 to 3 eV and some colors may be produced by ultraviolet radiation having an energy of several eV, particularly the 5 eV energy of short-wave ultraviolet (253.7 nm or 2537 Ångstrom units). Most color alterations require much more energetic rays, such as X-rays of about 10 000 eV or gamma rays of about 1 000 000 eV (or 1 MeV for mega electron volts).

X-Rays can be produced by the use of high voltages in *X-ray tubes* in relatively small machines, found in hospitals, dentists' offices, and industrial laboratories. The intensity available from these sources is usually rather low, so that long exposure times are required, which involve high electricity bills and the possibility of excessive wear and tear on the equipment. In addition, X-rays do not penetrate very deeply, so that the alteration they produce is usually located only on or just beneath the surface. Such X-ray-produced skin colors, therefore, are not usually employed for gemstone improvement.

Exposure to gamma rays in the laboratory is generally achieved in a gamma cell, such as the one shown in *Figure 4.2*. A gamma cell may be merely a large block of lead, weighing 3800 kilograms (8500 lbs) for the unit of *Figure 4.2*, and containing a central cavity. Within this cavity there are arranged rods, sheets, or pellets of a highly radioactive isotope, usually produced in a nuclear reactor, which emits gamma rays. A specimen to be irradiated is merely placed into the holder, seen open at the top of *Figure 4.2*, and is then lowered by an electric motor into the cavity for the desired exposure time. Other arrangements may be used.

Figure 4.2 A sample being loaded into a cobalt 60 gamma-irradiation unit by the author; when the doors are closed, the holder is lowered down into the cell. Photograph by courtesy of Bell Laboratories

The most commonly used isotope is cobalt 60, Co-60; that is, the isotope of cobalt having a nuclear mass of 60 units compared to ordinary Co-59; Co-60 emits two gamma rays of 1.17 MeV and 1.33 MeV. Cesium 137 (Cs-137, compared to ordinary Cs-133) is also used. There is no significant consumption of electrical power in a gamma cell, since the gamma rays are produced continuously. There is, however, a slow fall-off in the amount of available radiation: with Co-60 the intensity falls to one-half the original value in 5 years 110 days, to one-quarter the value in 10 years 220 days, to one-eighth the value in 15 years 330 days, and so on, in geometrical progression; with Cs-137 the time for each of these steps is 30 years. When the cell needs recharging, the whole unit must be shipped back to the supplier.

A typical Co-60 unit, such as that of *Figure 4.2*, provides an irradiation intensity of one Mrad (megarad = 1 million rads) per h, corresponding to 24 Mrad per day. In such a unit colorless quartz turns a deep smoky color in less than half-an-hour and saturates, that is achieves its darkest possible color, in less than one day; a slowly coloring Maxixe-type beryl may, however, require a month of irradiation to saturate. There exist large, commercial irradiation units, such as that shown in *Figure 4.3*, used for the polymerization of plastics, the sterilization of medical supplies, the preservation of food and the like, in which specimens a meter or more across may be moved by a conveyor, as in *Figure 4.3*, into the irradiation cavity. This is the size of a small room and may be so intensely radioactive that only minutes are required to achieve a dark-brown, smoky quartz.

Figure 4.3 The loading facility in a large, industrial, cobalt 60 installation; the irradiation occurs in the shielded room behind the left wall. Photograph by courtesy of IRT Corp

It should be noted that the designations of *Figure 4.1* have arbitrary boundaries. Thus, there is overlap and a high-energy X-ray of energy 500 keV is identical with a low-energy gamma ray of 500 keV. In the overlap regions the different designations refer rather to the way the radiation has been produced than to any difference in their natures.

Energetic Particles

The first energetic particles to be discovered were named before their nature was understood, so they are still sometimes called alpha rays and beta rays. The alpha particles or alpha rays are, in fact, energetic helium nuclei, while beta particles or beta rays, also called cathode rays or canal rays, are energetic electrons.

Electrons (light, negatively charged particles of unit charge) are usually produced by heating a filament (as in a vacuum tube or television tube) and are then accelerated by an electric field produced by applying a high voltage. A typical beta-ray energy is 1 MeV; one million volts is required to accelerate electrons to this energy, since this is the way the 'electron volt' energy unit is defined. High-energy electrons are most commonly produced in *linear* accelerators, powered by any of a number of different types of high-voltage generators, or in a *betatron*; a typical voltage-multiplier linear accelerator is shown in *Figures 4.4* and *4.5*. These units are large, expensive items and require considerable power and

Figure 4.4 A voltage-multiplier electron accelerator with the capability of producing 1.5 MeV electrons. Photograph by courtesy of Bell Laboratories

maintenance. Electrons that range into the billion eV range (abbreviated GeV for giga eV) can also be produced, but higher-energy electrons can induce radioactivity. Electrons are strongly absorbed and may yield only surface coloration at low energies. Much heat is generated in this process, so the material being irradiated is usually immersed in flowing cold water to minimize cracking. Electrical discharges can also occur.

The other charged particles used for irradiation, in order of complexity, are the proton, this being the positively charged hydrogen nucleus, H^+, having a unit charge and a mass of one atomic mass unit*; the deuteron, D^+, being the nucleus of

* One atomic mass unit is defined as one-twelfth the mass of an atom of carbon-12; it is equal to 1.66056×10^{-24} g.

the heavy isotope of hydrogen with unit positive charge and a mass of two atomic mass units; and alpha particles, He^{2+}, being the doubly charged nucleus of the helium atom with mass four; heavier ions can also be used. These particles are described in more detail in Appendix B. They are produced in a variety of very

Figure 4.5 Inside view of the electron accelerator of *Figure 4.4*. Photograph by courtesy of Bell Laboratories

large and complex machines, including *linear accelerators, cyclotrons, synchrotrons,* and the like, also described in Appendix B.; they can have energies ranging well above 1 MeV, but suffer the disadvantage that radioactivity may be induced.

Alpha particles of energy about 5 MeV are also emitted by *radium*, and early coloration experiments, particularly on diamonds, were performed by burying the specimens in a radium salt for some time. However, radioactive decay-products from the radium can become implanted into the diamond, leaving a serious radioactivity hazard in some stones.

Neutrons, uncharged particles of unit mass, are produced in huge quantities in a *nuclear reactor* or *pile.* The average energy of neutrons inside a reactor is quite low, but many neutrons have energies up to 8 MeV; those emerging from the reactor may be much less energetic, incorporating 'cold', 'thermal', and 'fast' neutrons with an average energy well below 1 MeV. The neutron flux is typically 10^{10} to 10^{14} neutrons per square centimeter per second. Neutrons are intensely penetrating; they can pass through matter readily because they have no charge and, since they are therefore not repelled by atoms, interact strongly when they do impinge on a nucleus. High energies are accordingly not required. Neutron coloration is thus a very uniform process but, unfortunately, neutrons may induce significant radioactivity.

Coloration by Irradiation

With the exception of diamond and the possible exception of some blue-turning topaz, the nature of the irradiation source is immaterial; the alteration is the same just as long as there is sufficient energy supplied by the irradiation – typically only 5 to 10 eV are needed. Of the various irradiation agents summarized in *Table 4.1*, gamma rays are clearly the preferred choice: they produce excellent uniformity of coloration, do not consume electrical power, do not produce localized heating, and do not induce radioactivity. The major irradiation-produced changes are summarized in *Table 4.2*.

Table 4.2 The major changes on irradiating gemstones

Material	*Changes*
Beryl and aquamarine	Colorless to yellow, blue to green, pale colors to deep blue 'Maxixe'[a]
Corundum	Colorless to yellow[b], pink to padparadscha[b]
Diamond	Colorless or pale colors to blue, green, black, yellow, brown, pink, or red
Pearl	Darken to gray, brown, 'blue', or 'black'
Quartz	Colorless, yellow, or pale green to smoky, amethyst, bicolor amethyst–citrine
Spodumene, kunzite	To yellow or green[a]
Topaz	Colorless to yellow[b], orange[b], brown[b], or blue
Tourmaline	Colorless or pale colors to yellow[b], brown[b], pink[b], red[b], or bicolor green–red[b]; blue to purple[b]
Zircon	Colorless to brown to reddish

[a] Color fades on exposure to light.

[b] Color may fade on exposure to light; there can be present in these materials at least two different color centers, one fading and one non-fading.

Many of the naturally occurring gemstones have been colored by irradiation over geological periods of time. This irradiation is derived predominantly from those radioactive elements in the earth's crust which were produced at the time of the creation of the universe, and still remain in significant quantities because of their exceptionally long decay periods. The most directly affected gemstone is zircon, which can contain in excess of 10 percent thorium and 2 percent uranium, both naturally occurring radioactive elements which, together with their decay products such as radium and radon, produce significant quantities of alpha- and beta-rays. This internally produced irradiation both colors the zircon and destroys the crystal structure, as described in Chapter 7.

Quite apart from such obvious radioactive sources there is also potassium, an element widely distributed in nature. Potassium is mostly K-39 and K-41, but also contains 0.012 percent of the isotope K-40 which emits beta- and gamma-rays. Several other naturally occurring elements also contribute small amounts of irradiation.

Within the top 2 m of soil, thorium, uranium, and their radioactive decay products produce, on average, an alpha-particle activity of about 2.4 disintegrations per square centimeter per minute, and potassium a beta-particle activity of about 9.5 disintegrations per square centimeter per minute. Although quite small, this activity is cumulative over long periods and, of course, varies considerably from place to place, being controlled by geochemical enrichment processes. Finally, there is the cosmic radiation, *see Figure 4.1*, which bombards everything on earth. The result of all this irradiation acting over geological periods of time has been to color much of the earth's topaz (blue or brown), its quartz (smoky, rose, or amethystine), its fluorite (purple or other colors), and so on. In some instances the material may have grown in colored form because of peculiarities of the growth conditions.

An important question then arises: why is not all topaz blue or brown, all quartz smoky, etc.? One reason may be that some of these materials were colored, but have lost their color from having been heated in nature sufficiently recently that there has not been enough time for the color to re-form. A second reason may be that in some localities the surrounding rock or soil is particularly free of radioactivity and acts as a shield. Finally, it may be that some of this material lacked either the electron- or hole-center precursors necessary for the formation of color centers, as described below.

Color Centers

The majority of the irradiation-induced color changes have not been studied in sufficient detail for a full understanding of the processes at work. Usually optical absorption-spectroscopy must be backed up by paramagnetic resonance and other sophisticated techniques to provide an unequivocal answer. However, even where such complete insight is lacking, the major outlines are clear. Most of these color changes involve a 'color center'. The characteristics of a color center include its production by irradiation (although other specialized techniques can sometimes be employed to produce the same result). Heating invariably produces the reverse change, although exposure to light or even merely sitting for some time in the dark may be sufficient to produce loss of color in some particularly unstable color centers. Invariably, a color center involves one electron missing from a normally occupied position, leading to a 'hole color-center', or the presence of one extra electron leading to an 'electron color-center'. If the electron is merely one involved in the variable valence of a transition element, then the term color center is not usually used. Some examples given below clarify these concepts.

Any material that can form a color center contains two types of precursors, which have the following characteristics. The 'hole-center precursor' is an atom, ion, molecule, impurity, or other defect which contains a pair of electrons, one of which can be ejected by energetic irradiation, leaving behind a 'hole center'. The 'electron-center precursor' is a similar atom, ion, etc., which can 'trap' the electron ejected from the hole-center precursor to produce an 'electron center'. The role of the irradiation is thus merely to transfer one electron, forming simultaneously a hole center and an electron center. Almost all materials have hole-center precursors; if there is no electron-center precursor, however, the electron displaced by irradiation immediately returns to its original place and the material remains unchanged.

If the initial hole-center precursor was originally neutral, designated as A in *Figure 4.6(a)*, then the hole center produced by irradiation with the loss of one electron is the positively charged A^+ of *Figure 4.6(b)*. Similarly, the neutral electron-center precursor B becomes the electron center B^- on acquiring one electron or 'trapping' it. Note that so far it has not been stated which of these four units, A, A^+, B or B^-, is the *color center* itself!

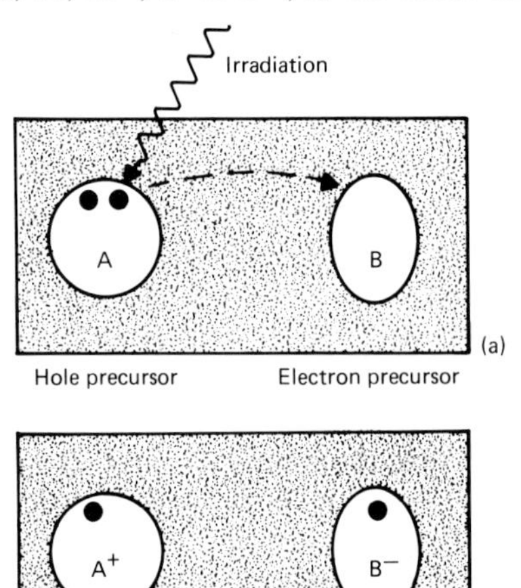

Figure 4.6 Changes on irradiation of a material that can form a color center: (a) before; (b) after. From K. Nassau, *The Physics and Chemistry of Color*, Wiley, New York (1983)

An unpaired electron can become excited by absorbing energy from a white-light beam, thereby removing part of the spectrum and producing color. Either or both of the hole- and electron-centers can do this. If it is only the hole center that absorbs visible light, a *hole color-center* is present; if it is only the electron center, an *electron color-center* is present.

If light or heat now liberates the trapped electron from the electron center B^- in *Figure 4.6(b)*, so that it can return to A^+ and convert it back into A, then the original state is restored and the color is lost, a process called fading or bleaching. If the trap is weak or 'shallow', even room temperature in the dark may be able to supply enough energy (about 0.1 eV) to release the trapped electron. With a trap a little stronger or 'deeper', it may require the 1 to 3 eV of visible light to produce fading, as in Maxixe beryl and unstable irradiated yellow sapphire or brown topaz. A deep trap may hold the electron so strongly that the alteration is perfectly stable to light, as in smoky quartz, amethyst, blue topaz, or stable brown topaz; heating to as high a temperature as 500 °C may then be required for bleaching. Such bleaching requires time; the higher the temperature, the more rapidly does it occur.

When irradiation brings about the change illustrated in *Figure 4.6*, it should be noted that the energy required is that to produce A^+ from A. When bleaching occurs, it is the change from B^- to B that is controlling. In a material that has one hole color-center precursor with several possible electron-trapping centers, the irradiation behavior is always the same, but the bleaching temperature may vary considerably, depending on which specific electron center is present in any given specimen. This explains, for example, why smoky quartz from different localities loses its color at temperatures ranging from 140 °C to 400 °C[1]. With an electron color-center, where it is the trap that provides the color, there can sometimes be several different hole centers, each requiring different energies to produce the color in specimens from different localities, and the coloration occurs at different rates in different specimens.

It is instructive to explore smoky quartz a little further at this point. Quartz is SiO_2, and the structure of a quartz crystal can be viewed as consisting of Si^{4+} and O^{2-} ions, all having only paired electrons in filled shells, as shown in *Figure 4.7(a)*.

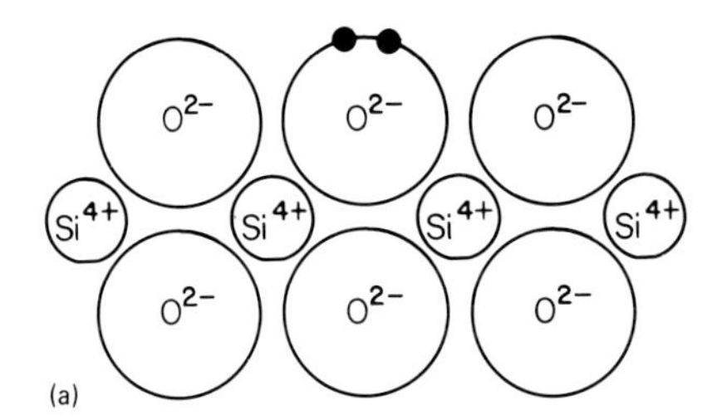

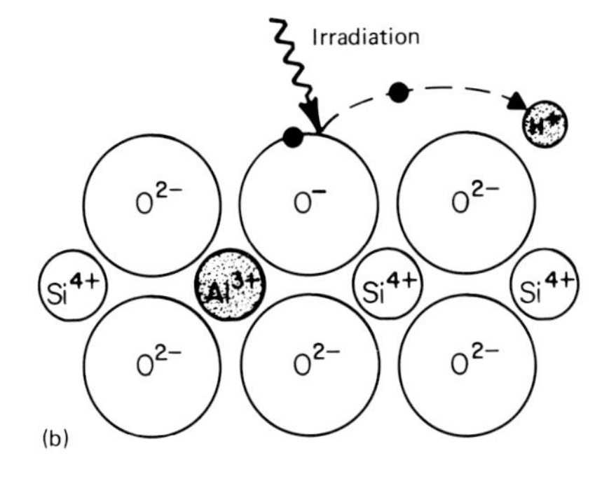

Figure 4.7 (a) Schematic representation of quartz and (b) the formation of smoky quartz by irradiation when aluminum and hydrogen ions are present. From K. Nassau, *The Physics and Chemistry of Color*, Wiley, New York (1983)

All quartz contains small amounts of aluminum, typically one Al for every 10 000 Si, as well as similar amounts of hydrogen and alkali-metal ions. When Al^{3+} replaces Si^{4+}, one positive charge is missing; the electrical balance can be restored by one H^+, Na^+, etc., for each Al present, located in some available open space in the quartz; an ion occupying such a position, which is normally vacant, is called an interstitial ion.

When irradiation ejects one of a pair of electrons, as shown in *Figure 4.7(b)*, it can be trapped at the H^+. This process can be written as:

$$[AlO_4]^{5-} \rightarrow [AlO_4]^{4-} + e^- \quad (4.1)$$

and

$$e^- + H^+ \rightarrow H \quad (4.2)$$

Here the hole center is viewed as a group containing the Al together with its four surrounding O, and the electron, shown as e^-, is produced by reaction (4.1) and trapped by reaction (4.2). The electron trapped by the H^+ converts it to a hydrogen atom, H. If Na^+ had been present, it would have formed a sodium atom, Na. The $[AlO_4]^{5-}$ of equation (4.1) is the hole color-center which gives the dark color to smoky quartz; bleaching is the result of the energy of heating reversing reaction (4.2) to release the electron, which then causes the reverse of reaction (4.1).

Smoky quartz is a good example of the complexity that can be found with color-center materials. Despite having been studied for many years, a recent re-examination of this problem by the author and B. E. Prescott[2–4] demonstrated that there are several color centers present in most smoky quartzes, in particular also a 'greenish yellow' quartz color-center of unknown structure, which can at times occur by itself. A combination of the usual smoky color-center with this as well as, possibly, other centers results in the wide range of smoky quartz colors, including greenish, reddish, and grayish tints.

The amethyst color-center works exactly like the smoky quartz one, with Fe^{3+} replacing Al^{3+} in *Figure 4.7* and in equation (4.1), except that now it is the $[FeO_4]^{4-}$ hole color-center that provides the color:

$$[FeO_4]^{5-} \rightarrow [FeO_4]^{4-} + e^- \qquad (4.3)$$

The electron once again is trapped by an electron-center precursor, as in equation (4.2).

On being heated, the reverse of equation (4.3) again produces $[FeO_4]^{5-}$, which provides only the pale yellow color of citrine (or 'burned amethyst' in part); this can once again be returned to amethyst by irradiation if it has not been overheated. This same mechanism also applies to pale yellow, synthetic quartz, which has been grown with iron added under oxidizing conditions so that Fe^{3+} occurs in place of Si^{4+}. If grown under reducing conditions, however, a green quartz results, containing Fe^{2+} in place of Si^{4+}. When this is irradiated, the Fe^{2+} is first converted into Fe^{3+}:

$$Fe^{2+} \rightarrow Fe^{3+} + e^- \qquad (4.4)$$

which can then form the amethyst color-center by the further absorption of irradiation, as in equation (4.3) above. Heating reverses the process to restore the green color, as also happens with some naturally occurring amethyst; the product is then called 'greened amethyst'.

The Restoration of Color

In some materials the same color center can exist in both fading and non-fading forms; one example is yellow sapphire. When a colorless sapphire is irradiated to become yellow, the color usually fades rapidly in the light. At the same time, non-fading yellow sapphire does occur naturally. This can present a difficult problem for the gemologist who is faced with such a stone, since in several such materials no test is known, at present, other than a fade test itself, to establish whether such a stone fades or not. Although both fading and non-fading color centers lose their color on heating, irradiation restores the material to its previous state, either fading or non-fading or even both!

On a number of occasions when accidental overheating of stones during jewelry repairing has removed the color from stable, red tourmalines and yellow sapphire, it has been possible to restore the colors by irradiation on the suggestion of the author[5]. It should be noted that the bleached yellow sapphire there reported turned a medium-dark, brownish yellow, but rapidly faded back to a 'pleasant 'Ceylon' yellow', a color which it had originally. From this description it is clear that the irradiation both produced the fading color center as well as restored the original stable color center; only the former was lost on exposure to light. A similar behavior could be expected with brown topaz, where both fading and non-fading color centers exist.

This type of behavior can be readily understood from the above explanation of color centers. If an electron color-center is involved, it means that there are two types of hole centers, one of which is deep and stable, while the other is shallow and fades. If the color is derived from a hole color-center, then there are two of these, one stable and one unstable.

The purple color of fluorite, the blue fading Maxixe and Maxixe-type beryl colors, and a number of other colors are also derived from color centers. These are discussed in the listing of individual gemstone materials of Chapter 7, where further references may be found.

The 'Stabilization' of Color Centers

Sometimes one encounters the suggestion that it might be possible to stabilize a color center by some suitable heat treatment. Based on the previous discussion it should be obvious that heating inevitably destroys a color center and cannot possibly preserve it. Heating may, of course, produce other changes, such as those described in Chapter 3, but that is quite another matter.

Color-center-like Color Changes

In the discussion of the greened amethyst above, it was noted that irradiation can change Fe^{2+} to Fe^{3+}, as in equation (4.4) for quartz. This change can also occur in beryl, where it is involved in the change of green aquamarine to blue aquamarine by heating. Green aquamarine contains iron in two different sites. When in an interstitial site within structural channels in the beryl, Fe^{2+} produces a blue color which is not affected by heat. Most specimens also contain some Fe^{3+} substituting for Al^{3+}; by itself this produces the yellow color of golden beryl or, together with the blue Fe^{2+}, gives green aquamarine. On heating, the Fe^{3+} in the channel can gain an electron released from an electron trap and then changes to Fe^{2+} by the reverse of equation (4.4); this substitutional Fe^{2+} contributes no color-causing light absorptions. Heat thus bleaches yellow beryl to colorless and converts green aquamarine into blue aquamarine. Irradiation can now return the yellow or the green color by reaction (4.4), as given above; heat can once again restore the colorless or blue state by the reverse of this equation, and so on. These changes, which do not involve color centres are further discussed under beryl in Chapter 7.

A similar change of color induced by irradiation which does not involve a color center is the deep green color produced by the irradiation of pink kunzite. This material contains manganese and iron, both in the trivalent state, Mn^{3+} and Fe^{3+}. Irradiation probably produces a coupled oxidation–reduction as follows:

$$Mn^{3+} \rightarrow Mn^{4+} + e^- \tag{4.5}$$

and

$$Fe^{3+} + e^- \rightarrow Fe^{2+} \tag{4.6}$$

The combination change is:

$$Mn^{3+} + Fe^{3+} \rightarrow Mn^{4+} + Fe^{2+} \tag{4.7}$$

Here the pink of Mn^{3+} is replaced by the deep green of Mn^{4+}. Light contains sufficient energy to reverse these equations and thus leads to a return of the color to pink. Other mechanisms may also be at work.

One last example is rose quartz, which appears to be colored by Ti^{3+}. Heating to 200–300 °C permits the change:

$$Ti^{3+} \rightarrow Ti^{4+} + e^- \tag{4.8}$$

to occur with a loss of the color, which can be restored by irradiation. The electron from reaction (4.8) is probably taken up by an iron impurity, as in reaction (4.6).

Radioactive Gemstones

Inevitably, we are all exposed to some radiation; the typical environmental radiation level is about 100 mrad (millirad) per year or 0.012 mrad per hour; these values can be halved or doubled, depending on the location and altitude (units used for radiation are described in Appendix B). In this exposure there are contributions from the radioactive components in our own bodies (over one million beta particles per minute, together with smaller amounts of alpha particles, etc.) as well as in the soil, from cosmic rays, and from fall-out of nuclear testing. Additional exposure is derived from medical and dental X-rays and from radiation treatments.

As mentioned above, zircon can contain radioactive elements, but the amount is usually so small in jewelry-grade material (*see* Chapter 7) that this can be ignored. Also discussed above and in *Table 4.1* is the fact that some treatments may leave the irradiated material radioactive. On several occasions, gemstones so treated were released into the trade. One example[6] was a parcel of 100 blue topaz stones, where one 10 carat stone showed 0.2 mr (milliroentgens) per hour on a survey meter, compared to a 0.02 to 0.05 mr per hour background reading. Another example[7] involved irradiated spodumene, where a 6 carat stone registered at 0.7 mr per hour. Much lower levels were found in some Maxixe-type blue beryl[8].

Gemstones having a significant radioactivity (the measurement of radioactivity is discussed in Appendix B) represent a totally unnecessary exposure; although there are no appropriate standards, stones such as these should not be used for extended personal wear according to a Nuclear Regulatory Commission spokesperson[6]. A dealer working with a package of many such stones would be at a

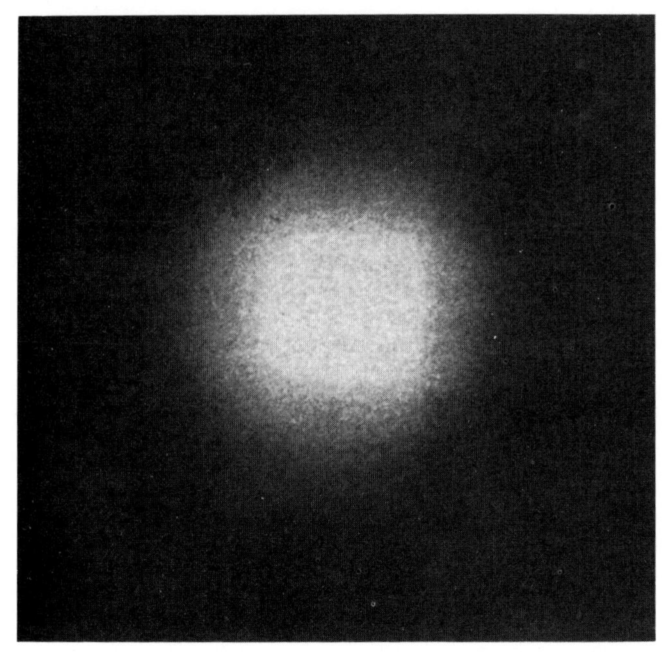

Figure 4.8 A photograph taken by placing the radioactive gemstone of *Figure 4.9* on a film overnight, also called an 'autoradiograph'

Figure 4.9 Lead container for storing a radioactive gemstone and the customary magenta-on-yellow radiation hazard sign

definite risk. Conscientious gemological laboratories now routinely check for radioactivity with a survey meter. Such radioactivity does decay with time, the exact rate of decay depending on the nature of the radioactive isotopes present.

So far such occurrences appear to have been isolated incidents. This problem may lessen, since it is reported[9] that the Brazilian authorities have been confiscating radioactive gemstones and investigating the unauthorized use of nuclear reactors.

Radioactivity is also seen in some old, radium-exposed diamonds. These stones may be colored or they may have had their color, but not their radioactivity, removed by a heat treatment. Although usually not too highly radioactive, at least one such stone was found to be exceptionally so, darkening film in as little as 15 minutes and giving the huge reading of 40 mr per hour on a survey meter[10]. Such an 'autoradiograph' is shown in *Figure 4.8*. When located, highly radioactive stones should be kept in a thick-walled lead container, equipped with the customary magenta-on-yellow radiation-hazard warning label, as shown in *Figure 4.9*.

References

1. K. Nassau, Irradiation-induced colors in gemstones, *Gems Gemol.*, **16,** 343 (1980); also, *Lap. J.*, **34,** 1688 (1980)
2. K. Nassau and B. E. Prescott, A reinterpretation of smoky quartz, *Phys. status solidi,* **A29,** 659 (1975)
3. K. Nassau and B. E. Prescott, Smoky, blue, greenish yellow, and other irradiation-related colors in quartz, *Min. Mag.*, **41,** 301 (1975)
4. K. Nassau and B. E. Prescott, Growth-induced radiation-developed pleochroic anisotropy in smoky quartz, *Amer. Min.*, **63,** 230 (1978)
5. C. F. Fryer (Ed.), Sapphire, color restoration, *Gems Gemol.*, **19,** 117 (1983)
6. R. Crowningshield, Irradiated topaz and radioactivity, *Gems Gemol.*, **17,** 215 (1981)
7. G. R. Rossman and Y. Qiu, Radioactive irradiated spodumene, *Gems Gemol.*, **18,** 87 (1982)
8. K. Nassau, The nature of the new Maxixe-type beryl, *Lap. J.*, **27,** 1032 (1973)
9. Anon, Brazil cracks down on nuclear gems, *Modern Jeweler,* **82,** 21 (Feb, 1983)
10. R. Crowningshield, Radium-treated diamond, *Gems Gemol.*, **12,** 304 (1968)

Chapter 5

Treatments other than Heating and Irradiation

Innumerable of stains and splendid dyes
John Keats

To dip his brush in dyes of Heaven
Sir Walter Scott

There are a wide variety of treatments in addition to those outlined in Chapters 3 and 4; these are covered in this chapter in three groups. First there are chemical treatments and impregnations which alter the bulk of a material, or at least penetrate into the surface. This includes bleaches, colorless oil, wax, and plastic impregnations, colored impregnations, and various forms of dyeing.

It should be noted that in Roman, Greek, and earlier times it was customary to paint or stain marble statues and other decorative objects with bright colors. Over the centuries most of this decoration has been lost or removed, so that our erroneous image of ancient objects is that of pure white marble. Even today, the staining of decorative objects such as jade carvings for their artistic enhancement is considered acceptable and is not necessarily intended to mislead.

In the second group there are surface coatings of various types, including surface paints, varnishes, and interference filters, as well as foil backs, surface decoration and inscribing, and the like. In the last group there are the composite stones which include doublets, triplets, gel- and insect-containing stones, as well as synthetic overgrowth, both of the same and of a different mineral. An abbreviated summary of the application of the enhancements covered in this chapter is given in *Table 5.1.*

Chemical Treatments and Impregnations

The simplest chemical treatment is cleaning, in any of its many forms. When soap or detergent and water, often used together with an ultrasonic cleaner, do not remove dirt or stains, dilute and then strong acids, particularly hydrochloric

Table 5.1 Enhancement processes covered in Chapter 5

Process	*Material*
Impregnations	
Bleaching	Chalcedony, coral, ivory, pearl, petrified wood, tiger's eye, etc.
Colorless oil/wax/plastic	Emerald, lapis lazuli, opal, ruby, sapphire, turquoise, etc.
Colored oil/wax/plastic	Beryl, lapis lazuli, quartz, ruby, turquoise, etc.
Dyeing	Agate, carnelian, chalcedony, coral, ivory, jade, onyx, opal, pearl, quartz, turquoise, etc.
Surface modifications	
Surface color coating	Amber, carnelian beads, diamond, pearl, etc.
Foil back, mirror back, star back	Used on any gemstone
Lasering	Diamond inclusions, gemstone identification
Glossing	Corundum, spinel
Composite gemstones	
Overgrowth	Emerald on beryl or quartz
Surface-diffused color or asterism	Ruby, various colored sapphires
Doublets, triplets, artifact-included, gel-filled	Amber, beryl, emerald, opal, ruby, sapphire, etc.

(muriatic) acid and ultimately aqua regia* are often used if the gem material can tolerate such corrosive agents. If yellow-to-brown iron stains or residues (rust) are a problem, a solution of oxalic acid (poisonous and corrosive) is often preferred. Care must be taken that a cleaning does not alter the color or etch the surface of a stone; it should always be tried out on a scrap first. Another danger can arise from the presence of a wax or oil, or a color treatment, which may be altered by too vigorous a cleaning. The same is true of organic solvents (acetone, methyl alcohol, ethyl alcohol, and so on) which are often employed to remove all traces of grease or oil before applying treatments such as heating or strong acids, which might otherwise produce dark and difficult to remove products.

Bleaching

Bleaching in everyday practice is performed with chlorine or hypochlorites, with hydrogen peroxide, with sulfites, or merely by the sun as in Shakespeare's

* Aqua regia is a mixture of 1 volume of concentrated nitric acid with 3 to 4 volumes of concentrated hydrochloric acid. This is a very dangerous mixture. It should be used only with good ventilation, since it gives off an orange–red gas, a mixture of the toxic and corrosive gases chlorine and nitrosyl chloride:

$$HNO_3 + 3HCl \rightarrow Cl_2 + NOCl + 2H_2O$$

Aqua regia should not be stored and the container should never, under any circumstances, be stoppered.

'The white sheet bleaching on the hedge'. All of these processes are also used for gemstones. Both diluted hydrogen peroxide and sunlight are employed to bleach pearls, natural and cultured, particularly dark ones and those showing a greenish shade, as well as coral; it is the organic *conchiolin* of these materials that is bleached in this way. Similar processes can be applied to other organic gem materials, such as coral to produce a golden shade from the black coral, and ivory to lighten material that has become excessively dark with age. Care must be taken in the latter case that drying out or cracking does not occur, for which reason ivory may be exposed to sunlight moderated by a sheet of glass.

Dick's Encyclopedia[1] gives several recipes for bleaching ivory, including sun exposure while wetted with soapy water (Section 2007, p. 200), rubbing with wet pumice and exposing to the sun under glass or using sulfurous acid, chloride of lime, or burning sulfur (Section 1997, p. 199), or placing the ivory into hot, thin lime paste (Section 1998, p. 199). Such procedures should never be tried without a preliminary test on a fragment, and even then their permanence cannot be assured.

Brown tiger's-eye is commercially bleached to give the more desired, paler 'honey' color. Two recipes for amateurs given by Zeitner[2] use 1 teaspoon of oxalic acid per cup of water applied for 12 h at 200 °F, or undiluted liquid bleach, such as 'chlorox', covered with a glass lid and placed in bright sunlight for as much as one week or even longer. Clearly, it is some of the iron color that is here being removed as in the cleaning discussed above.

Colorless Impregnations

There may be several different aims in applying a colorless wax, oil, or plastic to a gem material. First, the aim may be to hide cracks, as in 'oiling' an emerald or a ruby. Another technique aims to 'stabilize' a material such as turquoise to prevent acidic skin oils, for example, from entering the porous surface and producing a color change. At the same time, by a third mechanism, low-grade turquoise, which is white because of light scattering from its porous surface, may show a tremendous improvement in the color from an oil impregnation, which fills the pores and reduces the light scattering; the same process may convert worthless chalky opal into a brilliantly colored form, as shown in *Plate XXV*. Fourth, a hard coating over a roughly ground surface may make it look as if it has been polished. Fifth, and last, a colorless coating may be intended to provide a hard surface for a softer stone; a patent[3] was even issued for applying a 2 μm thick layer of sapphire over synthetic titania stones, although the author knows of no convincing evidence that any drastic improvement in hardness can be contributed by such a thin layer. On a number of occasions the growth of a thin layer of diamond over a diamond, and even over other surfaces, has been claimed, but so far no such claims have been convincing enough in the sense of supplying a useful growth rate to be generally accepted by the scientific community[4]. Old recipes for hardening stones, such as that of Monton cited in Chapter 2 on p. 19, do not really make any sense.

The ancient Romans used to wax their marble statues to hide small surface cracks and to give the surface a higher polish. The supposed origin of the word 'sincere' from the Latin *sine cera* for 'without wax' is, however, not correct. Mineral oil (liquid paraffin) is frequently used on the surface of carved objects to hide small imperfections; it is preferred to other oils because it does not dry out, is odorless, and does not turn rancid.

The oiling of emerald goes back to well before Pliny, who in 55 AD reported '*Smaragdi* [emeralds in part] . . . . in spite of their varied colours, seem to be green by nature, since they may be improved by being steeped in oil.' (Book 37, Ch 18, Vol 10, pp. 219, 221.)[5] Presumably white or brown-appearing, badly fractured stones are here improved by oiling.

Almost any type of oil, wax, or plastic can be used for treating emerald, ruby, or turquoise. Ordinary lubricating oil or the thicker types used for heavy machinery have been employed for emerald and ruby, but these have the disadvantage of tending to seep out of the cracks rather easily, particularly when the stones are warm. Paraffin wax is widely used for impregnating turquoise, usually with some heating, but plastic gives a much more permanent result. Recently, Canada balsam has come into wide use, particularly for emeralds, and so has the application of a vacuum, usually combined with gentle heating to perhaps 100 to 150 °C to improve penetration[6]. Although it is often said[7] that the impregnating agent is poured over the stones and then the vacuum is applied, a consideration of the physics involved indicates that it makes much more sense to apply the vacuum first and then add the oil under vacuum. (The hydrostatic head-pressure of the liquid exerts enough pressure to prevent small quantities of gas from escaping; heating helps in causing the gas to expand, but the hydrostatic head still remains.) A detailed description of the current emerald oiling process used in Colombia[7] has recently been given and is discussed in detail under Beryl in Chapter 7.

Canada balsam is the oleoresin now obtained from the North American balsam fir tree *Abies balsamea*; the equivalent substance was known in olden days as balsam of fir. It was already in use by 400 AD, as reported in the papyrus *P. Holm.*[8]; one example is given below. An advantage of Canada balsam over other oils and waxes is its high refractive index, which permits a better match with the refractive index of emerald than any of the other oiling substances listed in *Table 5.2*; as a result the cracks are less conspicuous. There are a number of disadvantages of oiling, apart from the obvious one of hiding defects. A stone setter can

Table 5.2 Refractive indices of various substances used at times for the oiling of gemstones

Material	*Refractive index*
Coconut oil, paraffin wax	1.45
Neat's-foot oil, palm oil, whale oil	1.46
Corn oil, mineral oil, olive oil, peanut oil, rapeseed oil, soybean oil	1.47
Castor oil, linseed oil	1.48
Light lubricating oil	1.49
Cedarwood oil	1.51
Tung oil	1.52
Canada balsam	1.53
Quartz	1.55
Beryl, emerald	1.58
Ruby, sapphire	1.77

easily break a stone, since he or she is not aware of its flaws. If the stone is heated, the oil or wax may char and produce irreversible damage. With time some oils, such as tung oil and other unsaturated oils, react with atmospheric oxygen and tend to dry-up, turning opaque. Cleaning, particularly ultrasonic, as well as some solvents such as a spilled alcoholic drink, also could remove the oil and unexpectedly demonstrate the previously hidden defects.

Colored Impregnations

Colored oil to fill cracks is used on a gemstone to improve its color, most frequently on emerald and ruby, or to make it look like another stone, most frequently applied to quartz. In an impregnation the solution of dye remains in the cracks, whereas in a dyeing the solvent evaporates and leaves the dye or pigment within pores, cracks or attached chemically to the dyed material. There are many recipes in the papyrus *P. Holm.*[8] which demonstrate that producing cracks in stones, if not already present, and then oiling them with colored oil was a common technique by 400 AD. The wide variety of mineral and biological substances mentioned in this papyrus has been described in Chapter 2, pages 10–12. One additional recipe[8] is surely justified for listing here:

> *Cooking of Stones*
> If you wish to make ruby from Crystal, which has been prepared for this purpose at your pleasure [i.e. pre-cracked], then take and place it in a *kerotakis* [a vessel usually used for melting wax with pigments] and stir in turpentine balsam [Canada balsam] and a little powdered alkanet [a red botanical dye], until the color sauce rises [bubbles?]. And then take care of the stone.
> (Author's translation.)

A perfect example of the colored oiling technique!

The late Robert Webster called the use of colored oil an 'unscrupulous trick', and there would seem to be little that can be said in favor of such a treatment. With the ready availability of organic dyes there is no difficulty in anyone obtaining supplies. Indeed, 'ruby oil' is widely available in Thailand[9] and 'emerald oil' in Colombia; some oils are shown in *Figure 5.1* . The label on the larger bottle of the ruby oil of *Figure 5.1* reads:

> *Crown Rubies Red Star*
> Perfectly increasing 100% value of gems, rubies and sapphiess [sic] with very shining and brightness. Soak your precious stones in the solution 'Crown Rubies Red Star' as long as required, then clean and polish with cloth.

Another bottle reads:

> *King Ruby Star Oil*
> Top chemical for cleaning especially ruby, garnet and fancy color stones. This chemical will help to increase the value up to 100%.

Depending on the dye used, such colored oiling may fade with time, in addition to suffering the other defects discussed under colorless oiling above. Fortunately, colored oiling can be relatively easily detected if the stone is immersed in a refractive-index matching fluid; this reveals the localization of the color. Rubbing with a cloth, cotton wool, or even white blotting paper dipped in a strong organic solvent may also reveal the colorant.

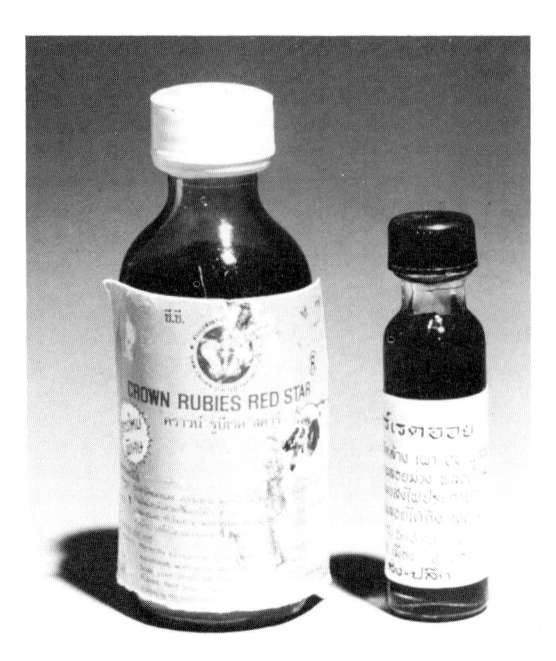

Figure 5.1 Oils used for enhancing the color of rubies in Thailand. Photograph by courtesy of the Gemological Institute of America

Dyeing Porous Chalcedony and Related Materials

Quite a large number of gemstone materials are sufficiently porous so that various dyeing or staining processes can be applied to them. Of these the most important comprise the cryptocrystalline quartz family, which includes chalcedony, agate, carnelian, onyx and the related materials listed under chalcedony in Chapter 7, as well as certain porous quartzites and the heat-cracked quartz described in Chapter 3. These all have the characteristic that they are porous, at least to some extent, are resistant to heat, and are also highly resistant to acids, so that a large variety of dyeing techniques can be applied. Materials other than these are discussed in the next section.

One of the simplest dyeing or staining processes involves concentrated hydrochloric (muriatic) acid. Most chalcedonies, agates, etc., contain small amounts of iron compounds such as iron oxide, frequently distributed as tiny inclusions. This iron can be dissolved under the action of the acid; if this is not washed out, but the stone next dried and heated, the iron becomes distributed as a thin film on all the internal surfaces and yields a yellow-to-brown-to-red color, depending on the temperature used. The heating aspects have already been discussed in Chapter 3. The addition of some iron salt or an iron nail or two to the acid will probably help to intensify the color!

A heat treatment at 200 to 300 °C has been reported to improve dyeing by opening up the pores of these materials in some special way[10], possibly by driving off strongly held surface water.

One of the best-known processes is the sugar or sugar–acid technique used to turn porous chalcedony deep yellow, brown, or black. The first mention of this process is found in Pliny (Book 37, Ch 74)[5]:

> Moreover, *Cochlides* or *shell stones* are now very common, but are really artificial rather than natural. In Arabia they are found as huge lumps and these are said to be boiled in honey without interruption for seven days and nights. Thus all earthy and other impurities are eliminated; and the lump, cleaned and purified, is divided into various shapes by clever craftsmen, who are careful to follow up the veins and elongated markings in such a way as to ensure the readiest sale. . . . . In general, all gems are rendered more colorful by being boiled thoroughly in honey, particularly if it is Corsican honey, **which is unsuitable for any other purpose owing to its acidity.**
> [Emphasis added.]

Ball[11] interpreted *cochlides* as being shell ornaments, but the 'huge lumps' description argues against this interpretation. Eichholz[5], on p. 323, more reasonably interprets these as large, inferior agates, boiled in acidic honey in order to bring out their color. Consider, now, the possibility of the use of an unmentioned, final, heating step at a slightly higher temperature, and one obtains something not too different from the modern sugar and heating or sugar–acid processes. Also note that Pliny's description of how the patterns of 'veins and elongated markings' are carefully followed is as true of clever craftsmen working with chalcedony and related materials today as it was almost 2000 years ago. This description could hardly apply to material derived from sea shells.

In a study published in 1847, 1848, and 1850, Noegerath[12] gives a detailed discussion. He points out that in expanding the application of this treatment to all stones in the last sentence cited, Pliny shows that he did not know the real mechanism. He suggests that sulfuric acid may have been known in Pliny's time and that a final heating step was used, exactly paralleling the modern process. However, because of a mistranslation from the Latin, earlier versions of Pliny did not mention the acid nature of Corsican honey, and one can reasonably conjecture that all that was required to char the acidic honey directly was a temperature somewhat higher than that used in the boiling, as is, indeed, still done today to obtain a brown color without the use of sulfuric acid, as described below.

Noegerath[12] gives details of the process as discovered (or rediscovered) about 1820 by an agate cutter in Oberstein and Idar, and subsequently brought into wide use. By breaking off a thin piece and touching this to the tongue, he reports that one can observe how different layers absorb the moisture and whether the material is suitable for producing onyx. After washing and drying at an elevated temperature, the material is placed into diluted honey and heated, but not to boiling point. After two to three weeks it is washed and placed into a pot of sulfuric acid covered with a mica plate, and heated for a few hours or a day on hot coals. After a final washing followed by drying in an oven, the material is shaped and polished and then placed for one day into oil to make small cracks disappear; any excess of oil is removed with bran. Pale gray bands can be turned a darker gray, brown, or black, depending on the porosity, while white non-porous regions are not affected; and so on.

Next, Noegerath mentions the production of yellow colors by the action of concentrated hydrochloric acid, and also the burning of agate, chalcedony, and carnelian at red heat to give various colors, but mostly reds. Lastly, he mentions a blue coloration process, still a trade secret at that time.

The history of the rise of the cutting and staining technology at what is now known as Idar Oberstein has been well documented. Among many others, there are personal accounts by Falz in 1926[13] and 1939[14], as well as the excellent series of ongoing articles by Frazier[15]. The sequence of activities included the beginning of heat treatment of carnelian in 1813, the sugar–acid process for black agate acquired from Italy in 1819, the yellow hydrochloric acid process in 1822, with later improvements, the ferrocyanide blue in 1845, and so on[15]. An early view of the town of Oberstein is shown in *Figure 5.2*.

Extremely interesting is the rare, small (vi + 20 pages) booklet *The Coloring of Agates*[16] published by Dreher in 1913; the cover and table of contents are shown

Figure 5.2 The town of Oberstein, from an engraving by Merian in 1645. Courtesy of S. Frazier

Dr. O. DREHER

Das

Färben des Achates

:: Idar 1913 ::

Verlag E. Keßler

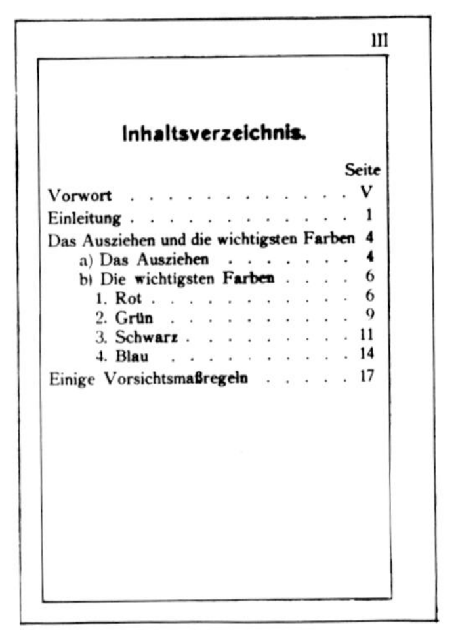

III

Inhaltsverzeichnis.

Figure 5.3 Title page and contents of Dreher's 1913 book on the techniques used for the coloring of agate in Idar Oberstein

in *Figure 5.3*. This was probably the first public disclosure of the 'secret processes', and he begins the foreword thus:

> No doubt many will disapprove that anyone in the field of our jewelry industry should reveal anything at all. One does speak so frequently of the danger of the copying of this activity elsewhere. However, I think that the following information dealing with generally known coloring methods contains nothing secret for the outsider. In addition, it is not possible for everyone to learn how to color merely by reading. Moreover, it does require a fair amount of experience. (Author's translation.)

It is obvious from the above that not all the tricks of the trade were being revealed! The avoidance of aniline dyes or other organic colorants, which frequently fade, was the real secret of these processes; instead inorganic coloring agents produced by chemical reaction or decomposition right within the pores were used, thus assuring permanence of color.

The process begins with cleaning the agate, first with soda solution to remove grease if necessary, and then by an acid treatment to remove traces of iron which would interfere with all colors, except where the red produced from this might not matter. Usually concentrated nitric acid is used, with the stones being soaked for 1–2 days and the temperature slowly rising to boiling point; the solution is cooled again before removing the stones. Three separate washes with cold water, each warmed and then cooled, complete the cleaning.

For red colors 0.25 kg of iron nails is dissolved with heating in 1 kg of concentrated nitric acid (toxic, corrosive, dangerous fumes). After being left to settle for several days, the consistency should be 'as thin as Munich Beer'. Stones are now soaked in this for 6–10 days if 3 mm thick, or 3–4 weeks if 10 mm thick. After careful drying at baking-oven temperature this soak is repeated, and the second drying (3–10 days) is continued directly up to red heat, when iron oxide forms within the pores. Drying, heating, and cooling must all be gradual to prevent cracking.

To obtain the green color of chromium oxide, two water-based dyeing solutions are used; the first saturated with chromic acid (toxic, corrosive), the second with ammonium carbonate. After soaking for 1–8 weeks in the first solution, the agate pieces are transferred to the second in a closed flat pan and left for at least 2 weeks; they are then dried and strongly heated. The color is not always clear, and there is mention of secret processes for improving it. A method using nickel nitrate and heating to produce a green color derived from nickel oxide is mentioned in passing.

Two liquids are also used for black agate, the first consisting of 375 g of sugar in 1 l of warm water, having the consistency of thinned honey. After being soaked in this for 2–3 weeks, the agates are transferred without drying into concentrated sulfuric acid (very dangerous). After being heated for 1 h and boiled for 15 min to 2 h and then cooled, the agates are removed, washed well, and then dried very well at medium temperatures. The concentrated sulfuric acid extracts water out of the sugar, $C_6H_{12}O_6$, leaving behind pure black carbon, C:

$$C_6H_{12}O_6 \rightarrow 6H_2O + 6C \qquad (5.1)$$

In a brief description, a brown color is stated to be produced by the action of heat on the sugar-impregnated materials; this color corresponds to the caramel color of candy-making.

Two techniques are given by Dreher for producing a blue color. The first uses a solution of 250 g of potassium ferricyanide (red prussate of potash – very toxic) in 1 l of warm water, in which the agates are soaked for 6–14 days. After being washed, the agates are transferred into a warm, saturated solution of ferrous sulfate containing a few drops each of sulfuric and nitric acids per liter. The deep blue color develops over 4–8 days. The second technique, only mentioned in passing, uses potassium ferrocyanide (yellow prussate of potash, also very toxic) and an iron salt, which would have to be a ferric salt such as ferric sulfate or chloride to produce the desired result. It has now been established that both of the products are exactly the same, variously known as Prussian blue, Berlin blue, or Turnbull's blue. The composition is most precisely described as iron(III) hexacyanoferrate(II), $Fe^{III}[Fe^{II}(CN)_6]_3$. This is an extremely stable pigment with its color derived from charge transfer involving the presence of both the ferrous and ferric valence states of iron.

Although most subsequent accounts of agate dyeing seem to be derived either directly or indirectly from Dreher, variants and other processes do exist. Cleaning may employ lye in addition to or instead of the soda. A black color is made by using a solution of cobalt nitrate and ammonium sulfocyanate (toxic) followed by careful heating, and another by using silver nitrate (toxic) solution followed by exposure to sunlight, which liberates metallic silver. Another process for a blue color uses a copper salt plus ammonia, although the blue product produced in recent years in Idar Oberstein employs a cobalt salt. Another process for a yellow color uses potassium dichromate (toxic), but without the use of heating; yet another uses a mercuric chloride (toxic) solution followed by potassium iodide to produce yellow mercuric iodide. These and many other variants have been assembled by Frazier[15], where further details and many additional references may be found. Some further comments are also given under chalcedony in Chapter 7.

Dyeing Materials other than Quartz

Most other gemstone substances which are dyed or stained can tolerate neither the strong acids nor the high temperatures used with many of the above quartz processes. Accordingly, natural dyes or aniline and other synthetic organic dyes are frequently used, even ordinary inks, marking inks, dyes used for fabrics, leather, wood, or even Easter eggs, cranberry juice, and so on! As might be expected, the resistance to fading of many of these dyes is far from satisfactory.

As discussed in Chapter 2, Pliny described the dyeing of amber with natural organic dyes such as alkanet or Tyrian purple and many subsequent reports have a variety of examples. Here is just one from Monton's *Secretos*[17], published in 1760:

> *How to Dye Marble Purple or Blue*
> Take the juice of black carrots and the juice of blue lillies. Since these two do not grow at the same time, try to save the juice of the one and mix it with the other when the time is right; also, you can do this with just one of the two. Have everything well strained and purified, boil it in white vinegar in equal parts, and for each pound of juice and vinegar add 1 oz powdered alum. Into this solution place the pieces of marble or alabaster and let them boil. Take them out when they are to your liking because the color gets darker the longer you boil it. And if the pieces are very large, you can brush it on, keeping both the marble and the dye hot.
> (Author's translation.)

Dick[1], about 1880, gives no less than twelve recipes for dyeing ivory, using silver nitrate and ammonium sulfide for black, cochineal for red, indigo or Prussian blue for blue, gold chloride for purple, verdigris for green, picric acid for yellow, and so on; he gives almost as many recipes for the dyeing of alabaster and marble[1]. Other substances that are frequently dyed in similar ways include coral, turquoise, jade, serpentine, various forms of calcite, pearls, and so on, as discussed in Chapter 7.

Identification of dyed materials may present problems. Careful microscopic examination usually reveals the coloration. Partial removal of the dye may occur when the material is immersed in or wiped with a white cloth soaked in alcohol, acetone, nail-polish remover, or the like. It is advisable to obtain permission before applying such a test, since it may well be destructive to the color.

Surface Modifications

Apart from oiling and dyeing, a wide variety of surface modifications have been practiced, including the application of wax, ink, surface coatings such as paints and varnishes, and interference filters, foil backs, mirror backs, inscribings, selective decorations, as well as synthetic overgrowth.

Water or saliva is often applied to a rough ground surface during the process of fashioning a gemstone to show the appearance of the final polished surface. A clear varnish or wax coating is frequently applied to poorly finished tumbled stones, cabochons, or carved objects to improve the appearance or even to avoid completely the polishing step.

The use of tints, that is paints or varnishes applied to the surface of a stone, was well described by Cellini[18] in 1568. It is most interesting that at that time in Italy the tinting of the surface of colored gemstones was forbidden by law, but the use of a colored backing in the cavity behind the stone was accepted, as was the application of colored coatings to diamonds. In Chapter 2, pp. 14 to 15, is described how Cellini achieved the most desirable smoky color by using gum mastic. The acceptable greenish colors for diamonds were produced by adding, for example, a green or a blue coating to a yellow stone. Today it is the colorless diamond that is most desired, but this too can be achieved for a yellow diamond by coating; most frequently, a small amount of purple ink is applied under the prongs of the setting or on the back of the stone. Purple is the complementary color to yellow, and in the process of 'physical bleaching' the effect of the diamond, which absorbs a little of all parts of the visible spectrum other than yellow, is balanced by the purple ink, which absorbs some of the yellow part of the spectrum. The net result is that only a little of the incoming white light is absorbed, without creating any impression of color[19]. Too much blue in such a treatment produces a bluish effect, not incompatible with the occasionally heard erroneous concept of the ideal 'blue–white' diamond color!

Bauer[20] describes how several different colors are sometimes applied to the back of a quartz or glass stone to imitate the prismatic color flashes characteristic of the diamond. More recently, this same type of effect has been achieved by applying thin *interference coatings*[19, 21], such as those found on camera lenses, with the color being produced by interference between rays of light reflected from the front and the back of the coating. When applied to glass rhinestones, the name 'Aurora Borealis Stone' has been used. Such coatings have been employed to modify the brilliance of zircon when used as a diamond imitation[22] or to increase the brilliance of other gemstones[23]. There is even a patent[21] for such a process.

In areas where the water supply contains a high iron concentration, it is possible for ring stones that have been frequently exposed to handwashing to acquire a thin, brown, tarnish-like, sometimes iridescent film of iron deposits. This is easily removed by gentle abrasion and is sometimes misinterpreted as an intentionally added surface coating[24].

Varnish or lacquer coatings have been used, at times for fraudulent purposes. A particularly noteworthy case[25] occurred early in 1983 when a 9.58 carat, potentially flawless, fancy pink diamond, valued at about $500 000, was being sold at an auction at the Sotheby Park Bernet galleries in New York. Someone successfully substituted a light yellow diamond of color U to V and clarity VS2 worth about a quarter of the original, which had been coated with pink nail polish; the deception was discovered just before the sale the following day.

Biringuccio[26] in 1540 noted that a black backing behind a diamond increases its brilliance; it must be remembered that diamonds were not faceted for total internal reflection in those days. This technique has also been used to hide dark patches in a stone by applying a dark mastic composition over the back of the lighter parts of the stone to even out the appearance, as described by Bauer[20].

The use of foil backs, shiny or colored metal sheets behind a gemstone to reflect light back toward the viewer was widely practiced in the days before faceting for total internal reflection. Even today it is sometimes used with cabochons, but

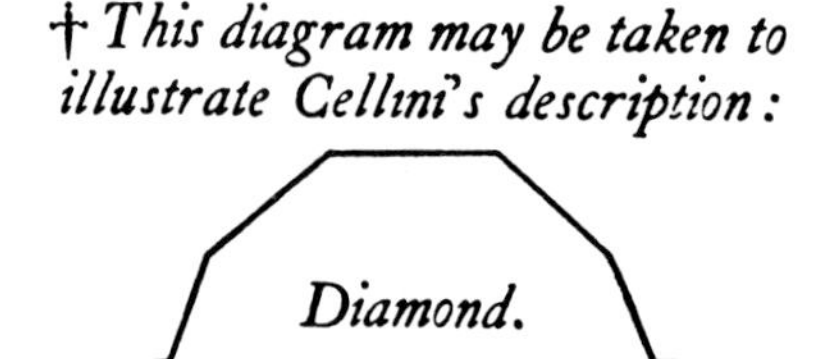

Figure 5.4 Reflector behind a diamond from the 1898 translation of Cellini's 1568 book on goldsmithing

most often to back glass stones for costume and stage jewelry. Extended descriptions were given by early writers including Pliny[5], Cellini[18], and Agricola as cited in Chapter 2, p. 16. A drawing from the 1898 translation of Cellini's 1568 book is reproduced as *Figure 5.4*. Both Porta[27] in 1658 and Dick[1] in 1880 (Section 2448 to Section 2459, p. 230) give 12 detailed foil recipes each. For colorless stones, highly polished colorless metals such as silver or tin were used, while colored metals or metals coated with colored varnishes were employed to intensify or modify the color of the stones. An example is given in the Cellini quotation on p. 15 in Chapter 2. There is also a technique which has been patented[28] in which a small hole is drilled into the stone at the culet and filled with colored material.

Related to foil backs is the practice of placing other colored substances behind a gemstone, including colored cloth, multicolored butterfly wings or peacock feathers, and so on; Bauer[20] gives a good summary. There is also a type of foil which is crimped to a very flat crown of diamond; this produces reflections as if there were a lower part to the stone. In this way a specific, thin, 10 carat, rose-cut diamond crown was made to look like a 50 carat solid stone[29]. Alternatively, a closed setting can be made to have the negative shape of the lower half of a faceted diamond, with the thin top set to seal the cavity and keep out dirt. Such a stone could be viewed as a diamond–air doublet!

The mirror back is a variant of the foil back; often colored, it produces a rather poor star effect. A somewhat improved version uses a metal foil that has been embossed or inscribed with scratches so that a cabochon placed over it appears to have a star in it. The scratches may be made directly on the back of the stone itself, a process that has been patented[30]; the scratches are usually applied in three sets at 120 degrees (i.e. 60 degrees) to each other. These processes can be applied to genuine or false composite stones.

Surface-modified stones include the surface-painted ones described above, carved, engraved, and etched stones, which are outside the scope of this book, as well as decorated beads. The latter, particularly those made of carnelian, have been found widely distributed in archaeological sites, as well as in Tibet, where those of a specific type are called Tsi beads. A detailed summary has been given by Frazier in Parts 9 and 10 of his series[15], where many references are cited. These beads are of several types; the most common have white geometrical patterns on the surface of carnelian or agate, while others may have dark patterns directly on the carnelian or on the whitened surface of the stone.

It is suspected by Frazier[15] that recent manufacturing is involved, since they are highly prized in Tibet where they are worn by all classes and where a wide range exists from obviously authentic to cheap imitations. Older processes used a slurry in water of sodium carbonate (e.g. washing soda), which is painted onto the carnelian and then dried and fired. By reaction with the silica this can produce a whitish, low-melting sodium silicate glass which resists weathering only poorly. A vegetable juice was sometimes added to the slurry to prevent flaking during the drying step. Other recipes incorporate a lead salt and potash, which would produce a similar low-melting, potassium lead silicate glass of similarly low weathering-resistance. Not all carnelian is suitable for this process, since some loses color during the heating step required to fuse the glass. Heating by burial in the embers of a hand-fanned charcoal fire was often used for this step, followed by a rather slow cooling.

The drilling of diamonds with a focused laser beam to burn out dark inclusions or make them accessible to a chemical treatment is an enhancement process that has been used widely in the last few years. An example is shown in *Figure 5.5*; this is further discussed under diamond in Chapter 7. In a 'new high in potential deception'[31], cubic zirconia has been laser-drilled to make it more convincing as a diamond imitation.

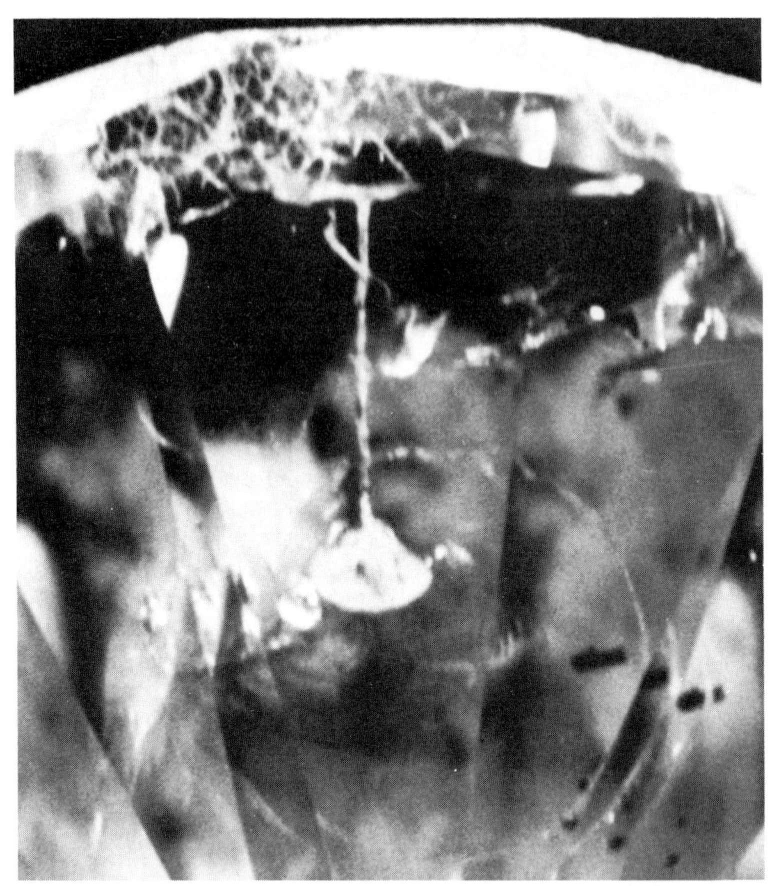

Figure 5.5 A diamond showing a laser-drilled hole. Photograph by courtesy of the Gemological Institute of America

Figure 5.6 A diamond with its certificate number lasered on the girdle. Photograph by courtesy the Gemological Institute of America

The inscription on the girdle of a gemstone of a certificate or registration number, typically 10 μm tall, with a focused laser beam, has recently been introduced by the Gemological Institute of America. An example is shown in *Figure 5.6*.

Although a discussion of conventional grinding and polishing is not here considered a part of surface modification, a brief mention may be made of fire polishing or 'glossing' the surface of a gemstone material, as described in two patents[32]. By applying a flame to fuse a thin layer of the surface of, for example, corundum or spinel, resolidification can produce a very smooth surface. According to the patents, this process can also be performed after applying a thin coating of a lower melting substance; presumably this could then form a glass, glaze, or enamel layer on the surface.

Somewhat related is the softening of ivory with phosphoric acid so that its shape can be altered, as described under ivory in Chapter 7.

Composite Gemstones

In Chapter 2, p. 7, is given a quotation from Pliny showing how the ancient Romans used to glue together three differently colored stones, one black, one white, and one red, to produce sardonyx, which would presumably then be used for the manufacture of cameos or intaglios.

There are many different types of composite or assembled gemstones. It is customary to speak of a doublet if two pieces of either the same or different materials are combined, usually with a colorless cement, while if three pieces are used, or if two pieces are assembled with a colored cement, then the designation triplet is employed, as illustrated in *Figure 5.7*. There are some minor variations in terminology in different countries. Webster[33] devotes more than 12 pages to the many types of doublets and triplets and should be consulted by those interested; a much briefer outline is here presented. The use of composite stones peaked early in this century and declined rapidly with the rise of inexpensive synthetics, such as the Verneuil products[4]. Nevertheless, they are still seen, particularly as emerald imitations, where a low-cost synthetic is not available.

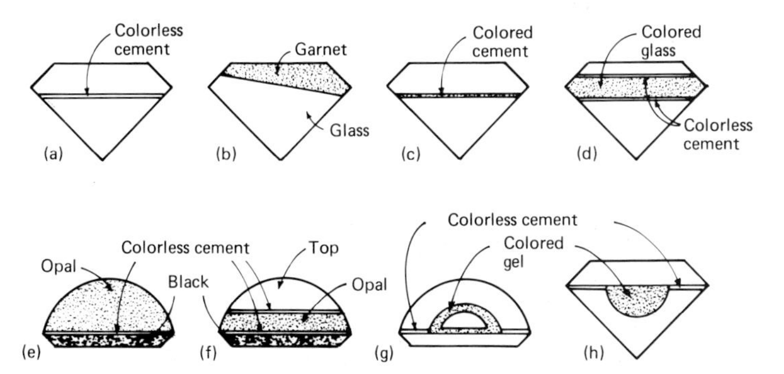

Figure 5.7 Different forms of doublets and triplets: (a) doublet; (b) garnet-top doublet; (c) triplet; (d) triplet; (e) opal doublet; (f) opal triplet; (g) + (h) gel-filled triplets

In a 'true' or 'genuine' doublet* two pieces of the same material are cemented together. Examples include two flat diamonds assembled to make one of normal faceted shape (rarely done in the past because of the lack of a good, non-deteriorating cement for diamond); two pieces of emerald, ruby, or sapphire, only one of which need have a good color, as in a beryl–emerald doublet; and the frequently seen opal doublets, where a piece of precious opal may be backed by a piece of common or 'potch' opal, which may even be dyed black. For cameo doublets the above-mentioned technique described by Pliny has been used.

In *semi-genuine doublets* there is a crown (or upper part of a cabochon) of the genuine material backed by a pavilion (or lower part of a cabochon) of another substance. Examples arc diamond, ruby, sapphire, or emerald backed with quartz, natural or synthetic spinel, or glass; opal backed with black onyx, black glass, etc., as shown in *Figure 5.7* (the opal may even be used in the form of small mosaic pieces); beryl or pale emerald backed with green glass[34], and so on. A recent example is a natural ruby–synthetic ruby assembly[35], while another involved a synthetic blue spinel cemented to a flat base of natural star sapphire[36], with the asterism clearly visible. Also falling into this group are the 'diamond–air' doublets discussed above, as well as rose quartz backed by a blue-colored mirror to intensify the asterism[30].

In *false doublets* the material being imitated is totally absent; there are many possibilities here, including rock-crystal over green glass[34]; the *nifty gem* combinations of a soft, but brilliant, strontium titanate pavilion covered by a much harder crown of sapphire in a *laser gem* or of spinel in a *carnegie gem*; an imitation opal made of rock crystal or glass over a slice of abalone shell; a turquoise imitation made of a suitably colored glass cemented over a dyed chalcedony back; fibrous ulexite 'television stone' in a cabochon triplet to imitate a cat's eye; and garnet-top doublets. These last are based on a discovery, made about 1850, that molten glass can be made to adhere to natural garnet. Such doublets were made in large quantities some years ago, most frequently employing a thin slice of almandine garnet backed by variously colored flint glass made to imitate emerald, ruby, sapphire, topaz, peridot, amethyst, and even diamond. Since the garnet is rather thin, its red color is not normally seen against the color of the glass. The garnet slices, often cut on a slant in the hurry of commercial production, as shown in *Figure 5.7*, were placed into cavities in clay molds, a piece of glass was added, and the whole was then placed into a furnace to melt the glass. After cooling the garnet-topped doublets were faceted.

There are even two relatively recent patents on the design of doublets[37, 38].

The variety of triplets is as great as that of doublets. The type used for cameos and intaglios reported by Pliny is mentioned at the beginning of the section. There may be a layer of colored gelatin in a 'genuine', 'semi-genuine', or 'false' triplet, or the third component may be a layer of colored glass, as also shown in *Figure 5.7*. The best-known triplets are the *soude* stones, named after the French word for 'soldered'. These were originally made of two pieces of rock crystal 'soldered' together with green-colored gelatin to give *emeraude soude*, as well as in other

* Composite-gemstone terms such as 'genuine', 'semi-genuine', and so on are not much used today; they do, however, serve as a useful pedagogical device in organizing the subject matter.

colors to imitate other gemstones. In the 1920s they were replaced by similar stones using a thin layer of colored glass with a colorless cement holding the three pieces together. With the ready availability of colorless, synthetic spinel, this material later replaced the quartz in all types of triplets. In a variant of this[34], a quite thick, colored glass layer has been used as also shown in *Figure 5.7.*

Much more difficult to recognize because they have the correct inclusions are the 'genuine' triplets such as *mascot emerald* and *smaryll*, made of two pieces of aquamarine or poorly colored emerald held together with a deep green cement. The equivalent has also been done with colorless or pale sapphire and ruby. In opal triplets, also known as *triplex opal*, a protective cap of rock crystal is added to an opal doublet, as shown in *Figure 5.7.* Also used have been caps of synthetic colorless spinel and even synthetic colorless sapphire. An imitation star that is made of a synthetic corundum triplet containing a diffraction grating layer in the center has been patented[39]. A rather poor opal imitation has been made from a colorful piece of mother-of-pearl in a false triplet.

The use of a gel or 'tincture' to fill a cavity within a stone to improve the color is mentioned by Leonardus in 1502, as cited in Chapter 2, p. 14; examples are given in *Figure 5.7.* For a short period, starting about 1958, this has been used to improve the color of jade in 'imperial jade'-colored, gel-filled triplets made from translucent white jadeite. Moss agate has been imitated in a triplet in which a dendritic shape is formed by diffusing chemicals into a thin layer of gelatin, which is then dried and cemented into a triplet[33].

Yet another type of composite or assembled stone is made with amber, where insects, fishes, and other objects have been encapsulated within amber, as described under amber in Chapter 7.

Composite or assembled gemstones can be recognized readily if the joint is visible, for example, by a change in the luster of the surface. They are, however, frequently mounted so that the joints are hidden in the setting. Observing a discontinuity in appearance as one focuses with a microscope down through the stone often reveals the nature, as do bubbles or swirl marks in the cement. Immersion in an index-matching fluid, or even in water, can frequently reveal the localization of the color both in composite as well as in dyed stones. The spectrum of any dye used can also be diagnostic. Garnet-top doublets in colors other than red show a red ring when placed table down on a white surface. Opal doublets may present a particular difficulty, since natural opal frequently occurs in thin seams which may have the same appearance as a doublet. The lack of dichroism in a gemstone where it should be present can be an indication of an artificial coloration.

Synthetic Overgrowth

About 1960, Johann Lechleitner of Innsbruck, Austria, began to market an emerald consisting of a pale beryl preform on the surface of which a thin layer of dark green beryl had been grown by a hydrothermal technique. The names *emerita* and *symerald* were at times used for such a product. Details of this and other synthesis procedures have been given[4]. The surfaces of these stones frequently show many small cracks, presumably derived from a difference in the properties of the natural stone and the synthetic overgrowth layer. The properties of this surface layer have been described by Bank[40], who also raised the question whether these stones should not be called 'synthetic emerald–beryl doublets'. In at least one instance[41] a layer of synthetic emerald was similarly grown over a natural quartz preform, probably accidentally.

Many years earlier, between 1916 and 1928, Professor Nacken in Germany had grown synthetic emeralds, at first thought to be made by a hydrothermal process. A detailed examination by the author (ref. 4, pp. 131–140) demonstrated that these were, indeed, grown from the flux, but on the surface of fragments of pale natural beryls.

Natural ruby has also been used at times as a seed crystal (*see*, for example, ref. 4, pp. 49–51), and the author has collected some synthetic, hydrothermal ruby specimens which were also grown over natural seeds. One of the first successful quartz growth attempts by Spezia in 1908 produced synthetic overgrowth on a natural quartz crystal (ref. 4, pp. 100–101). All in all, it seems likely that almost any synthetic process might be successfully used to produce overgrowth. It is not clear whether such a product should be called a natural–synthetic doublet, a synthetic coating, or merely a synthetic.

The surface-diffused corundum with color or asterism added, as discussed in Chapter 3, can also be considered to fall into the composite gemstone category.

References

1. W. B. Dick, *Dick's Encyclopedia of Practical Receipts and Processes or How They Did It in the 1870s,* Funk and Wagnall, New York (1974)
2. J. C. Zeitner, The mail bag, *Lap. J.*, **36,** 1464 (1982)
3. S. E. Mayer, US Patent 3 539 379, Nov 10 (1970)
4. K. Nassau, *Gems Made by Man,* pp. 197–200, Chilton, Radnor, PA (1980)
5. D. E. Eichholz, *Pliny: Natural History,* Vol 10, Harvard University Press, Cambridge, MA (1962)
6. G. O. Wild, Tränken von Schmucksteine im Vakuum, *Z. dt. Gemmol. Ges.*, **23,** 321 (1974)
7. R. Ringsrud, Emerald treatment in Bogota, Colombia, *Gems Gemol.*, **19,** 149 (1983)
8. O. Lagercrantz, *Papyrus Graecus Holmiensis (P. Holm.), Recepte fur Silber, Steine und Purpur,* Uppsala (1913)
9. P. C. Keller, The Chanthaburi-Trat gem field, Thailand, *Gems Gemol.*, **18,** 186 (1982)
10. H. C. Dake, Heat treatment agate coloring, *Mineralogist* (Portland, Oregon), **18,** 100 (1950)
11. S. H. Ball, *A Roman Book on Precious Stones,* pp. 193–194, 324, Gemological Institute of America, Los Angeles, CA (1950)
12. Noegerath, Die Kunst, Onyxe, Carneole, Chalcedone und andere verwandte Steinarten zu faerben, zur Erlaeuterung einer Stelle des Plinius Secundus, *Bonner Jahresb., Vereins v. Altertumsfr. im Rheinlande,* **10,** 82 (1847); also *Archiv f. Mineralogie, Geognosie, Bergbau u. Huettenkunde,* **22,** 262 (1848); also *Edinb. New Phil. J.*, **48,** 166 (1850)
13. E. Falz, *Die Idar-Obersteiner Schmuckstein-Industrien,* Carl Schmidt, Idar (1926)
14. E. Falz, *Von Menschen und edlen Steinen,* E. Falz, Idar (1939)
15. S. Frazier, The gem cutting industry of Idar Oberstein, Parts 1–8, *United Lapidary Wholesalers Show News,* **2**(2), 4 (1978); **2**(3), 6 (1978); **3**(2), 6 (1979); **4**(1), 14, 44 (1980); **4**(2), 14, 23 (1980); **4**(3), 12, 27 (1980); **5**(1), 14, 32 (1981); **5**(2), 10, 24 (1981); Parts 9–14, *Jeweler/Lapidary Business,* **5**(3), 20, 38 (1981); **5**(4), 24, 60 (1981); **6**(2), 20, 50 (1982); **6**(3), 24 (1982); **6**(4), 24, 60 (1982); **6**(5), 8, 52 (1982); Parts 15 on, *Jeweler/Gem Business,* **7**(1), 19, 50 (1983); **7**(2), 20 (1983); **7**(3), 18 (1983); **7**(4), 9 (1983)
16. O. Dreher, *Das Färben des Achates,* E. Kessler, Idar (1913)
17. B. Monton, *Secretos de Artes Liberales y Mecanicas,* Madrid (1760)
18. B. Cellini, *The Treatises of Benvenuto Cellini on Goldsmithing and Sculpture,* Dover Publications, New York (1967)

19. K. Nassau, *The Physics and Chemistry of Color: The Fifteen Causes of Color,* Wiley, New York (1983)
20. M. Bauer, *Precious Stones,* Vol 1, pp. 90–91, Dover Publications, New York, 2 Vols (1968)
21. H. S. Jones, US Patent 3 490 250, Jan 20 (1970)
22. L. B. Benson, New color alteration fraud becoming common in jewelry trade, *Gems Gemol.,* **6,** 79 (1948)
23. A. Tremayne, Improved gem brilliancy by special coatings claimed, *Gemmologist,* **18,** 73 (1949)
24. R. Crowningshield, Diamond discolored by water, *Gems Gemol,* **12,** 22 (1966)
25. C. F. Fryer, Painted pink diamond: The big switch, *Gems Gemol.,* **19,** 112 (1983)
26. V. Biringuccio, *Pirotechnia,* p. 122, MIT Press, Cambridge, MA (1942; paperback edition, 1966)
27. J. B. Porta, *Natural Magic,* pp. 186–189, Basic Books, New York (1957)
28. G. B. Kitchel, US Patent 3 835 665, Apr 13 (1973)
29. R. Crowningshield, Foiled again!!, *Gems Gemol.,* **13,** 374 (1971–1972)
30. R. S. Mukai, US Patent 2 511 510, Jun 13 (1950)
31. C. F. Fryer, Diamond simulant, *Gems Gemol.,* **19,** 172 (1983)
32. M. H. Barnes and E. L. McCandless, US Patent 2 448 511, Sep 7 (1948); US Patent 2 608 031, Aug 26 (1952)
33. R. Webster, *Gems: Their Sources, Descriptions and Identification,* 4th Edn, pp. 456–468, revised by B. W. Anderson, Butterworths, London (1983)
34. R. Crowningshield, Rare doublets, *Gem Gemol.,* **12,** 307 (1968)
35. C. F. Fryer (Ed.), Ruby and synthetic ruby assembled stone, *Gems Gemol.,* **19,** 48 (1983)
36. R. Crowningshield, Unusual doublet, *Gems Gemol.,* **12,** 305 (1968)
37. H. S. Jones, US Patent 3 755 025, Aug 28 (1973)
38. H. S. Jones, US Patent 3 808 836, May 7 (1974)
39. J. G. Donadio, US Patent 3 088 194, May 7 (1963)
40. H. Bank, Aus der Untersuchungspraxis, *Z. dt. Gemmol. Ges.,* **29,** 197 (1980); **30,** 118 (1981)
41. H. Bank, Aus der Untersuchungspraxis, *Z. dt. Gemmol. Ges.,* **28,** 166 (1979)

Chapter 6

The Identification of Treated Gems and the Question of Disclosure

Detection is, or ought to be, an exact
science and should be treated in
the same cold and unemotional manner.
Sir Arthur Conan Doyle

The brief outline of gem-testing techniques given in this chapter is not intended to enable the inexperienced reader to perform his or her own treatment identifications. Even if this chapter were to be book-length, and even if a reader were to study carefully the standard texts such as Anderson's *Gem Testing*[1] and Liddicoat's *Handbook of Gem Identification*[2], backed by Webster's *Gems*[3] and the author's *Gems Made by Man*[4], this would still be insufficient. What is required is that ultimate of gemological instruments, the experienced eye, with the informed mind to direct the investigation. Such experience is best gained through one of the hands-on gemstone-identification courses, such as those given by the Gemological Institute of America, Santa Monica, California, and New York City, New York, and the Gemmological Association of Great Britain, London.

Only by knowing the full range of possible enhancing treatments as listed in the next chapter can all the tell-tale signs be interpreted to lead to the best answer. At the same time it must be recognized that this 'best' answer is not necessarily the 'correct' one in the sense that there are certain treatments which leave no evidence of their use or which would require unacceptable destructive tests for their certain identification.

The aim in this brief outline is to provide the non-gemologist reader with an idea of the general and special gemological techniques used in the identification of treatments.

There are official guidelines, both in the US and the European Economic Community, which apply to the jewelry and gemstone trades. These guides establish the regulatory background for the disclosure of treatments, a subject discussed at the end of this chapter.

Gemstone Testing

One of the serious limitations on gemological identification is the inability to perform destructive tests – one can hardly blame the owner of a valuable gemstone for wishing it to retain its full worth. (If all else fails, the gemologist's report may indeed only be able to contain the statement 'origin of color undetermined'.) The research gemologist who is involved in developing the required non-destructive tests has no such constraints and frequently does destroy, intentionally or even unintentionally, the specimens. Nevertheless, certain 'slightly' destructive tests are occasionally performed when the situation requires it.

The problem of the detailed examination of a gemstone in a totally enclosed setting is more tractable, in that careful removal and resetting of a gemstone does not normally produce damage. Certain tests, with the color grading of a diamond being a prime example, just cannot be carried out with a stone in its setting. In the field of enhancement, it is the use of color behind a stone and the presence of difficult composite stones that is most troublesome in this respect.

In approaching an unknown gemstone, the obvious first question is the nature of the material, making due allowance for the possibility of composite, coated, and dyed stones. With the nature established, the next step concerns the possibility that one of many treatments, such as those described in Chapter 7 under the specific material at hand, may have been used. Both of these determinations may require the use of the full range of gemological tests.

Over the years a number of instruments have been developed to assist the visual and tactile senses in gemstone testing; a number of these are indispensable to the gemologist.

Visual inspection can reveal many of a gemstone's secrets under favorable conditions. A thorough but gentle cleaning is an essential first step in any examination. The color, as seen by the naked eye, gives surprisingly little information since so many gemstones occur in a wide variety of colors (such as sapphire, spinel, tourmaline, quartz), or colorants may have been used. The luster, indicative of the refractive index, fire caused by the dispersion, birefringence (double refraction), and dichroism (pleochroism) all can be evaluated approximately by the trained eye. Special optical effects such as asterism are also noted.

Evidence of surface coatings, external junctions, of cement layers in doublets, characteristic fracture (cleavage, conchoidal, and so forth), any filling in cracks, and the presence of characteristic inclusions and other growth features, all can provide diagnostic data.

Magnification is most important for this, either in the form of a simple 10× *loupe*, or as one of the sophisticated *microscopes*, possibly equipped with provisions for binocular viewing and dark-field illumination, as in the instrument of *Figure 6.1*. The type and distribution of imperfections and growth irregularities are the most important characteristics examined. By the ability to focus down through a stone, it is possible to detect boundaries such as those in doublets or triplets, as well as surface coatings. A photography attachment can help with documenting the observations, although the information is rarely as detailed as that observed by the eye. The use of a refractive-index matching fluid permits the observation of color localization. Immersion in any fluid is helpful; water (RI = 1.33) frequently serves,

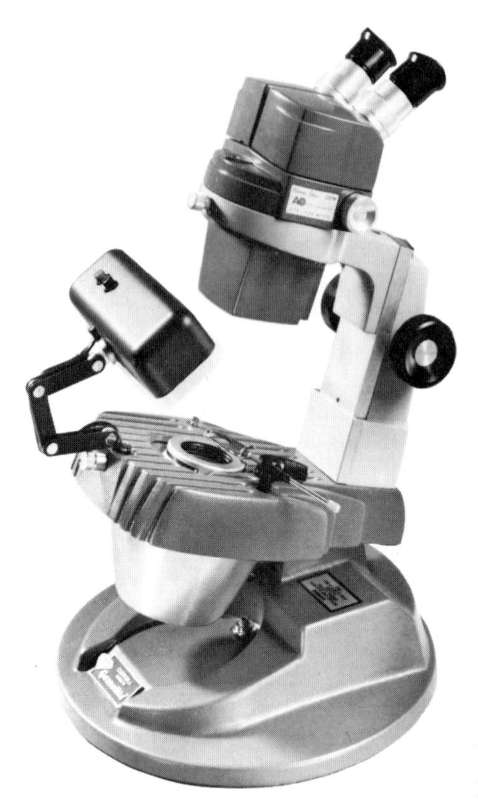

Figure 6.1 The GEM Custom 'A' Mark VI Gemolite binocular microscope. Photograph by courtesy of the Gem Instruments Corp

as does a saturated salt solution (RI about 1.38). Olive oil or glycerin with an RI of 1.47 often give much better results. Methylene iodide at 1.74 is very useful, but should not be used on porous or easily stained materials.

The *refractometer*, such as that of *Figure 6.2*, together with a *polariscope* and a *dichroscope* (even a simple piece of a sheet polarizer serves for both of these in experienced hands) permit the determination of the refractive indices, the optical character, the presence of double refraction or anomalous strain, and the pleochroism. If the refractive index is higher than the limit of the refractometer scale, a *reflectometer*, such as that of *Figure 6.3*, may give useful information. A *thermal probe*, such as that of *Figure 6.4*, can readily distinguish diamond from other gemstones and may also help in other identifications. It should be recognized, however, that the instruments described in this paragraph usually only probe the surface layer, so that their output must be treated with care if a coated, overgrown, or a composite doublet or triplet stone may be at hand.

The *spectroscope*, usually equipped with a source of bright light as in the unit of *Figure 6.5*, is most useful, both for identifying the material and showing the presence of dyes and coatings from their characteristic spectra. The fluorescence under *ultraviolet light*, both under short-wave (254 nm) and long-wave (366 nm) radiation, and the transparency and fluorescence under *X-rays* can be helpful, but allowance must be made for the wide variability in fluorescence for different localities, possibly derived from the presence of small amounts of unusual

Figure 6.2 The GEM Duplex II Refractometer. Photograph by courtesy of the Gem Instruments Corp

Figure 6.3 The Hanneman Diamond Eye, a reflectivity meter. Photograph by courtesy of Hanneman Gemological Instruments

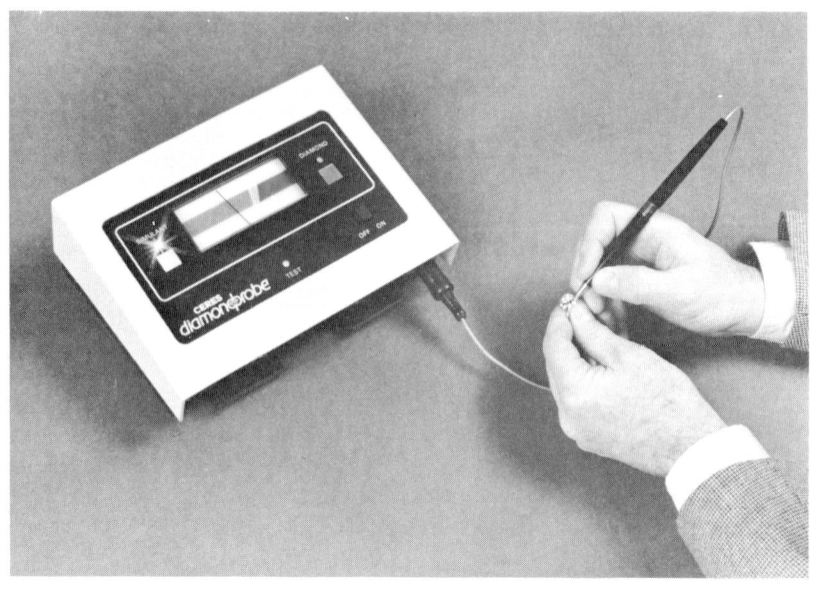

Figure 6.4 The Ceres Diamond Probe, a thermal probe. Photograph by courtesy of the Ceres Corp

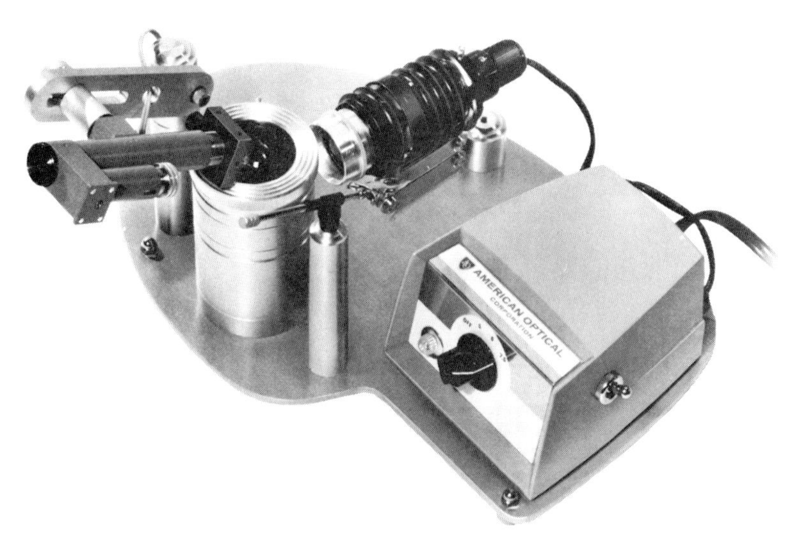

Figure 6.5 The GEM Spectroscope Unit. Photograph by courtesy of the Gem Instruments Corp

impurities. Some of the oils used in the impregnation of emeralds and rubies show a fluorescence. The identification by *X-ray diffraction* needs to be used only rarely. Specialized *filters*, such as the 'Chelsea' or 'emerald' filter, rarely give unambiguous results, but can provide additional clues in a difficult identification.

The specific gravity or density as determined by flotation or weighing in air and in water can be a useful test for the identification of loose stones. The hardness is rarely checked, except in unusual circumstances on an inconspicuous spot, if available, or by gently testing the stone for its ability to scratch a reference specimen.

Specialized instruments include a *conductometer*, which measures the electrical conductivity and helps establish the origin of the blue color of a diamond (conducting if natural), and a *hot point*, an electrically heated needle, carefully applied to the back of a cabochon or an inconspicuous part of a carving, which can sometimes reveal the wax or plastic impregnation of turquoise; the odor can even identify the impregnating compound. Similarly, amber, more recent resins, and plastic imitations may be distinguished. The careful use of acids can distinguish calcite, coral, and other carbonates from other materials. Hydrochloric acid with lapis lazuli gives hydrogen sulfide, with its 'rotten egg' odor. The use of ether for the identification of amber is discussed under amber in Chapter 7. Solvent may sometimes remove color from a dyed object, as described in Chapter 5.

A portable testing kit is available in the GIA *GEM Mini-Lab*, shown in *Figure 6.6.*

Figure 6.6 The GEM Mini-Lab. Photograph by courtesy of the Gem Instruments Corp

When evidence for the enhancement of a gemstone has been found, the gemologist will wish to known the answers to a number of important questions:

(1) Does the enhancement treatment correspond either partially or completely to an equivalent natural process?
(2) Can the presence of the enhancement be identified by a non-destructive or, at worst, a 'slightly destructive' gemological examination?
(3) Is the color stable or does it change with time under wear or display conditions?

The detailed listings of Chapter 7 attempt to give answers to these three questions. Some of the answers are also required in deciding on the necessity for disclosure, as discussed in the next section.

Many of the treatment processes are the exact equivalent of natural processes that occur during the formation of gemstones or their subsequent history. In a number of the natural color centers, the color was almost certainly produced by the exposure over geological periods to the low-level radioactivity in the environment, as described in Chapter 4; amethyst and smoky quartz are examples. In a few instances these processes are not directly equivalent. Thus, some of the naturally occurring dark-blue Maxixe beryl turns green rather than blue after fading and irradiation; an additional factor is clearly involved. In other instances one may suspect that the natural color centers may have been produced during a specific growth process, even though the apparently identical color can also be produced by irradiation. Examples of this latter may be blue topaz, where extremely high dosages are required, as well as the above-mentioned Maxixe beryl.

There are instances where the heat treatment is equivalent, as in some sun-bleached greenish yellow quartzes, greened amethyst, or in asteriated corundum. Yet here too the conditions may have been quite different. Thus in the last example, stars can be developed in some natural corundum by heating it to temperatures significantly higher than those likely to have been present for most naturally occurring asteriated corundum. The reason is that time is required for diffusion of the star-forming titanium. This diffusion can occur in a few hours at 1300 °C in the furnace or over many centuries at much lower temperatures in nature. This difference also provides one of the possibilities of detecting the use of a furnace treatment, since the very high temperature may leave evidence of its use, such as the cracking around inclusions discussed in Chapter 3.

The question of 'slightly destructive' testing, particularly for fading colors in those instances where there is no indirect test, is not clear. Certainly a number of tests in common gemological use are not, in fact, totally non-destructive. One immediately thinks of the hot-point probe, ether applied to amber, solvents to black pearls, and the minute scrapings taken from the girdle of a stone for an X-ray diffraction test in a really difficult identification. It may well be that it is now time for a fade test to become a routine procedure, as proposed some years ago by Crowningshield. As discussed below, this presents additional problems because some natural gemstones may fade (some brown topaz and some kunzite, for example). The resolution of such a problem clearly requires a decision by the jewelry trade as a whole.

The Disclosure Regulations

With the positive identification of a treated gemstone the question arises: to what extent should, or must, an enhancing treatment be disclosed when material is being sold? This is the subject of the Federal Trade Commission of the United States (FTC) guides[5,6] and the International Confederation of Jewel(le)ry*, Silverware, Diamonds, Pearls, and (Precious)* Stones (CIBJO) guides[7]; the latter apply to the European Economic Community and other participating countries. Only those aspects having relevance to treatments are here discussed.

* The spelling and wording varies within the document!

In the US, the proposed modification[6] of the FTC guides[5] which, in a modified form, is soon expected to be accepted by the FTC, provides a set of disclosure guidelines (the version current in late 1983 is used; changes may still be underway).

Speaking for the US via the Jewelers' Vigilance Committee, the recommended revision[6] states in paragraph 23.1(a):

> It is an unfair trade practice to sell . . . . any . . . . product . . . . having the capacity and tendency or effect of deceiving as to the type, kind, grade, quality, color, durability, serviceability, value . . . .

Under diamond, there is paragraph 23.11:

> It is recommended that lasering be disclosed. . . . . any . . . . diamond weighing 0.20 carats or more, when . . . . can be detected . . . . magnified at ten power . . . .

On gemstones, paragraph 23.20(c) states:

> The following are prohibited: The sale or offering for sale, of any diamond or other gemstone, which has been artificially colored or tinted by coating, irradiating, or heating, or by use of nuclear bombardment, or by any other means, without disclosure of the fact that such gemstone is artificially colored, and disclosure that such artificial coloring or tinting is not permanent, if such is the fact.

There follows an important exception:

> Note. – Treatment need not be disclosed if either one of the following conditions applies: (1) Heat treatment not detectable by a qualified gemological laboratory, or (2) the results do not revert under normal wear or display conditions. (Extreme caution should be exercised in such a non-disclosure, it being expressly understood disclosure would avoid confusion and liability which could attach from failure to disclose.)

In a similar vein, speaking primarily for the European Economic Community, CIBJO[7] states in Art. 5 that:

> a) Stones which are coloured or improved in colour by physical treatment, chemical treatment or physical–chemical treatment:–
>
> I) whose colour has been altered by irradiation
> II) whose colour has been altered by chemical treatment
> III) which have been coated
>
> MUST be designated as TREATED and the mineral name of the variety must be used.
>
> c) Excluded are the gemstones and ornamental stones in the following list, which have undergone a permanent and irreversible colour transformation by thermal treatment or by the effect of acid only:
>
> Agate (veined agate–cornelian–onyx–green agate–blue agate)
> Beryl (aquamarine–morganite)
> Quartz (citrine–prasiolite)
> Topaz (pink topaz)
> Tourmaline (all colours)
> Zoisite (blue tanzanite)
> Corundum.

On pearls, Art. 20 specifies:

> a) All artificial coloration of natural pearls is to be stated clearly and directly
> b) Regarding cultured pearls, slight alterations of tint (e.g. rosetint) due to treatment which they have undergone, do not need to be specified.
> c) On the other hand, all artificial colorations, e.g. grey or black, are to be clearly and directly specified . . . .

Although these excerpts from the guides seem to be quite specific, a close reading indicates uncertainty in application. As one example, in CIBJO Art. 20 above, at what point does a 'slight alteration of tint' change to an 'artificial coloration'? And how can a gray irradiation-produced color be 'directly specified' if it cannot always be detected with certainty? Similarly, the exceptions to FTC paragraph 23.20 present uncertainty: does the lack of detectability of treating have to apply to a specific stone or to all such material, and what if it is detectable only some of the time? Neither guide specifically addresses the question of the heat-treatment induced diffusion of color- and star-forming impurities into the surface of a sapphire, presumably because this was not a problem at the time of writing.

The CIBJO[7] guide uses (as did an early version of the proposed FTC guide) the technically quite precise term 'reversibility' when it appears that stability in the sense of reverting under wear and display conditions was intended. The intentional reversal of treatment changes is often possible with irradiation or heat-treatment processes, quite unrelated to the question of the stability.

The FTC guides are perhaps best summed up by two principles: the avoidance of deception *and* the approval of a treatment without the necessity of disclosure *either* if use of the treatment cannot be detected *or* if its results are stable. An obvious exception (not specifically mentioned at present) to the alternative option in the last sentence is diamond, where both conditions are usually considered to be necessary for the avoidance of disclosure.

Disclosure in Practice

The listing of *Table 6.1* is an attempt to clarify the disclosure situation by considering eleven specific treatments. These were selected from the many discussed in this book because they span the full range from the heated aquamarine of item A, a treatment hardly ever absent and not in the least likely to produce any deception, to the aquamarine or pale emerald impregnated with a deep-green colored oil of item K, which is surely intended to deceive and which any ethical seller would disclose, were he or she to handle such material. There are four heating processes (items A to D), three irradiation colors, one stable (E) and two unstable (F and G), three impregnations, two colorless (H and J) and one colored (K) and, finally, the sapphire colored or asteriated by impurity diffusion (L).

The first five treatments, items A to E, are almost never disclosed. This seems consistent with the FTC guides since here there is no possibility of deception; in addition, it is not usually possible to identify the treated products with the certainty that would be required for the enforcement of disclosure, even if it should be required. Of the three irradiation-induced colors, the blue topaz of item E has a stable color, is not gemologically identifiable, and accordingly is quite appropriately not usually disclosed. Item G, the irradiation-produced Maxixe-type deep-blue beryl color which fades in light is hardly practical for wear; this is readily identified by gemological testing and again would be disclosed by the ethical dealer, were he or she to handle it.

Item F, those irradiation-produced colors that fade but for which there is no gemological test known (at present), poses a difficult problem for the trade. The only way the gemologist can identify this treatment, so that disclosure could be enforced, is by a fade test which, however, destroys the color; Crowningshield long

Table 6.1 Summary of some typical treatments, their identification, and their current disclosure in the US gemstone and jewelry trade

	Example of material used and treatment techniques	*Is product stable?*	*Is treatment detectable by routine testing?*	*Is treatment disclosed in US trade?*
(A)	Aquamarine turned from green to blue by heat	Yes	No	No
(B)	Zircon heated to turn colorless or blue	Virtually all	No, but these colors very rare in nature	No
(C)	Sapphire or ruby heated to remove silk	Yes	Sometimes	No
(D)	Sapphire heated to modify or develop a blue color	Yes	Usually, but not always	Sometimes
(E)	Topaz turned blue by irradiation	Yes	No	No
(F)	Topaz or sapphire irradiated to a yellow or brown color	No	No, only fact of fading	Usually
(G)	Beryl irradiated to produce a Maxixe-type blue color	No	Yes	Yes
(H)	Turquoise or opal impregnated with a colorless stabilizer	Usually	Usually, but not alway	Sometimes
(J)	Emerald or ruby impregnated with a colorless substance	Variable	Usually, but not always	Sometimes
(K)	Beryl or emerald impregnated or coated with color	No	Yes	Yes
(L)	Sapphire impurity diffused to produce a surface color or surface asterism	Yes/No[a]	Yes	Yes

[a] Note that this color or asterism lies in the surface and may be removed during recutting or repolishing.

ago foresaw the possibility that a fade test might one day become necessary[8]. There is the further complication that some natural material, such as some brown topaz in particular, also fades, so that the fade test in itself does not demonstrate treatment as such. Nevertheless, it could usually be so viewed for the simple reason that the color of fading, natural brown topaz would not normally survive the collection, selection, faceting, setting, display, and sales sequence; indeed, such material has rarely been seen in the trade.

There appears to be unanimous agreement in the US trade that the last two items of *Table 6.1*, the colored oiling of item K and the surface diffusion of item L, should be disclosed since they would otherwise represent deception. Bank[9] views an emerald overgrowth on natural beryl, discussed in Chapter 5, pp. 72–78, as a synthetic emerald–beryl *doublet*; one could view the surface-diffusion induced color or asterism in the same way.

Item H, 'stabilized' turquoise is disclosed some of the time but not consistently. It could truly be claimed that here the 'enhancement' is real rather than merely an effort at deception, since the properly treated material has a color which is stable to acidic skin oils and the like, as discussed in Chapter 5. At the same time, it is also true that apparently good-color turquoise can be made by impregnating rather poor white material, as also discussed there.

At the time of writing, most disagreement seems to be centered on the question of the disclosure of colorless oiling, particularly of emerald, discussed in Chapter 5. There is the difficulty that gemological examination cannot always identify this treatment; accordingly some claim that oiling should never need to be disclosed, while others maintain the opposite point of view. This topic has been the subject of considerable discussion at a number of recent symposia panels on which the author has served[10,11], as well as in recent trade publications[12,13]. The conclusion of these discussions seems to be that both national and international agreements would be most desirable.

References

1. B. W. Anderson, *Gem Testing*, 9th Edn, Butterworths, London (1980)
2. R. T. Liddicoat, Jr, *Handbook of Gem Identification*, 11th Edn, Gemological Institute of America, Santa Monica, CA (1981)
3. R. Webster, *Gems: Their Sources, Descriptions and Identification*, 4th Edn, revised by B. W. Anderson, Butterworths, London (1983)
4. K. Nassau, *Gems Made by Man*, Chilton, Radnor, PA (1980)
5. Federal Trade Commission of the United States, *Trade Practice Rules for the Jewelry Industry*, promulgated June 28, 1957, amended November 17, 1959 and February 27, 1979, as *Guides for the Jewelry Industry, 16 CFR, Part 23*, Washington DC
6. W. S. Preston and J. A. Windman, *Recommendations for Revisions of: Guides for the Jewelry Industry, 16 CFR, Part 23*, presented before the Federal Trade Commission of the United States, Jewelers' Vigilance Committee, Inc, New York (1981)
7. CIBJO (International Confederation of Jewel(le)ry, Silverware, Diamonds, Pearls, and (Precious) Stones), *Gemstones – Pearls*, British Jewellery and Giftware Federation, London (1982)
8. R. Crowningshield, Developments, *Gems Gemol.*, **9**, 294 (1959)
9. H. Bank, Aus der Untersuchungspraxis, *Z. dt. Gemmol. Ges.*, **30**, 118 (1981)
10. I. Z. Eliezri, Panel Chairperson, *Legitimacy of Treatments, Changes in Gemstones and the Need for Disclosure*, International Precious Stones Congress, Tel Aviv, Israel (Apr 13, 1983)
11. E. Rosen, Panel Moderator, *Disclosure of Gem Enhancement: To Tell or Not to Tell*, International Society of Appraisers, Beltsville, MD (May 24, 1983)
12. H. Huffer (Ed.), Treated gems: Why not the facts? *Special Projects Report*, Jewelers' Circular Keystone, 30–80 (May, 1983)
13. M. Ford, Disclose stone treating? GIA to take hard look at hot gem biz issue, *National Jeweler*, 1,20 (May 16, 1983)

Chapter 7

Specific Gemstone Treatments

I'll swear her color is natural:
I have seen it come and go.
Richard Brinsley Sheridan

Would not detect therein one circumstance
To show a change from what it was before.
Henry Wadsworth Longfellow

Introduction

Treated materials are discussed under their own name, not under that of the material being imitated. Thus, chalcedony dyed to imitate jade is discussed only under chalcedony, but dyed jade is discussed under jade.

Designations in the right-hand margin, such as **A**, **B**, **C**, . . ., are used for the various processes given in the summary table for each material or group of materials of any complexity, and these designations are repeated in the margin wherever each process is discussed. Note that a specific treatment may be discussed in more than one place within any section. The marginal symbol # indicates the gemological identification discussion.

Abbreviations used in the summary tables are as follows:

[s] – stable
[u] – unstable *or* can be unstable
[f] – fading *or* can be fading
[r] – reversible by another treatment process

It is assumed that the reader has read through Chapters 3 to 6; there is much important and relevant information in these chapters that is necessary for the full understanding of the material discussed below. As mentioned previously, the descriptions of treatments given here should not be viewed as recommendations. The venturesome reader should always first try any procedure on a small fragment of little value.

For background information on the specific gemstones, Webster's *Gems*[1] is highly recommended and is referred to as 'Webster' in this Chapter, without any further designation. The *Encyclopedia of Minerals*[2] is a very useful reference manual, as is the *Jewelers' Dictionary*[3]. Bauer's *Precious Stones*[4], although 80 years old, is still an invaluable reference. The identification of the use of a treatment from the results of a gemological examination has been discussed briefly in Chapter 6. The gemology texts by Anderson[5] and Liddicoat[6] should be consulted for detailed guidance, as well as the manuals for the GIA gemological courses on colored stones and on gem identification, and the courses of the Gemmological Association of Great Britain (*see* Appendix D).

There are a number of early reports of heat- and irradiation-treatments which are not included below; these either involve very minor changes or those that have not been confirmed by later workers, often because incorrectly identified materials had been used. The fact that a treatment was attempted and did not work is not listed; after all, the raw material may just not have been suitable for the process tried.

The absence of a listing may mean that there is no such treatment or that it does exist and either has not been reported or merely that the author has not found it.

References

1. R. Webster, *Gems: Their Sources, Descriptions and Identification,* 4th Edn, revised by B. W. Anderson, Butterworths, London (1983)
2. W. L. Roberts, G. R. Rapp, Jr, and J. Weber, *Encyclopedia of Minerals,* Van Nostrand, New York (1974)
3. D. S. McNeil (Ed.), *Jewelers' Dictionary,* 3rd Edn, Chilton, Radnor, PA (1976)
4. M. Bauer, *Precious Stones,* Dover Publications, New York, 2 Vols (1968)
5. B. W. Anderson, *Gem Testing,* 9th Edn, Butterworths, London (1980)
6. R. T. Liddicoat, Jr, *Handbook of Gem Identification,* 11th Edn, Gemological Institute of America, Santa Monica, CA (1981)

Alphabetical Listing

Abalone, *see* pearl

Achroite, *see* tourmaline

Adularia, *see* feldspar

Agate, *see* chalcedony

Agatized Bone, Coral, Wood, etc., *see* chalcedony

Alabaster, *see* gypsum

Albite, *see* feldspar

Amazonite, *see* feldspar

Amber

Including: gedanite and succinite

Summary of Treatments

Heat:
- **A** Darkening [s]
- **B** Clarification [s]
- **C** 'Sun-spangle' cracks [s]

Heat plus pressure:
- **D** Reconstruction [s or u]

Other:
- **E** Dyeing [s or u, f]
- **F** Coating, foil, composites, etc. [s or u, f]

Many of these amber treatments can also be applied to the more recent natural resins, such as copal resin and kauri gum.

Amber is the time-hardened resin exuded by certain ancient pine trees. It is a mixture of organic substances, mostly terpenes, with a composition near $C_{10}H_{16}O$; succinic acid, $HOOCCH_2CH_2COOH$ is a significant ingredient but is present at a concentration of less than 10 percent. Amber softens over a broad temperature range from 150 to 200 °C and melts at 250 to 300 °C. Detailed descriptions have been given recently by Vávra[1], Rice[2], and Webster; older, valuable reports include those of Dahms[3], Williamson[4], and Schmid[5].

A Some amber will darken by itself, even in the dark[6], a process probably involving air-oxidation. Heating has long been used on pale amber to produce darkened or 'antiqued' material; this makes it look more like the preferred 'age oxidized' amber. Williamson[4] describes a long, slow heating by burying the amber in sand in an iron pot; Schmid[5] also gives many details. The conditions used vary with the composition of the amber. **A**

Amber that contains tiny gas bubbles may be merely turbid, may resemble goose-fat ('fatty', 'flohmig' in German), or may be quite cloudy ('bastard') as described by Webster. These bubbles may be less than 1 micrometer in diameter (0.00004 in) and there may be as many as 1 million per square millimeter visible in a cut surface. Clarification, the process of removing such bubbles from amber, has been discussed in Chapter 3, pp. 42–44. This can employ a slow heating in rapeseed oil[3], linseed oil, or fat as described by Pliny[7] almost 2000 years ago, or even without any fluid medium, merely in the dry state[3]. Several days may be required, since both heating and cooling must be performed extremely slowly to prevent cracking. **B**

This cracking may be desirable in itself, although one suspects that the original popularity of the cracked product may have derived from a good publicity campaign to dispose of amber damaged during unsuccessful clarification attempts! The cracks usually result from uneven heating and occur in the form of 'nasturtium leaves', round disk-like shapes as illustrated in *Figure 3.9* and *Plate XXI*; they are usually called 'stress figures' or 'sun-spangles'. The product is stable as long as it is not exposed to excessive heat. **C**

Reconstruction, the process of using heat and pressure to combine small pieces of amber into large ones, has been described in Chapter 3, pp. 42–44. If the pieces are carefully selected, clear, and trimmed to be free of dirt, the product can be difficult to recognize as being reconstructed. This process appears to have been first used in 1880 in Vienna and rapidly became a commercial success. Careful examination can usually, but not always, reveal flow lines and elongated bubbles; an unusually bright blue fluorescence may also be noted. Some pressed amber may turn white with time, possibly because the process was not performed properly; otherwise the product is stable. **D**

Dyeing amber to produce darker and more desirable colors has been widely reported. Pliny[7] was quoted in Chapter 2, p. 7, on the dyeing of amber with natural dyes such as alkanet and Tyrian purple and its use as an imitation of other gemstones. Williamson[4] mentions a Chinese dyeing treatment that employs pressure to produce a deep-red cherry or plum color. Being an organic substance, amber can be expected to interact well with organic dyes, but fading in light could be a problem if the dye is not chosen carefully. Dye or other coloring matter can also be added during both clarification and reconstruction processes. **E** **B** **D**

Coatings, such as varnish, to give amber a darker color have been reported on a number of occasions[8]. Foil at the back of amber to reflect light was much in vogue at one time, as described in Chapter 5, but is rarely seen today; colored foils were also used to modify the color. It is sometimes assumed that the presence of an insect in amber is a guarantee of authenticity. However, Vávra[1], Rice[2], and Webster mention and Andrée[9] gives a detailed report on composite stones in which insects and other objects (even fish!) have been encapsulated by using an embedding resin or plastic within pieces of amber, copal, or kauri gum; these have been produced for hundreds (or even thousands) of years. **F**

Dick[10] has this recipe for joining pieces of amber: **F**

2176. To Cement Amber
Amber is joined or mended by smearing the surfaces with boiled linseed oil, and strongly pressing them together, at the same time holding them over a charcoal fire or heating them in any other way that will not injure the amber.
(p. 211.)

One suspects that today an epoxy resin would do a safer and better job. Amber may be diluted with dammar resin and it is widely imitated by a variety of plastics. Amber chips have also been embedded in plastic[11].

Identification is based on a microscopic examination and gemological constants; these need to be backed up at times by a hot-point test, a check of the solubility in ether, and the application of a knife blade. #

Natural amber may be almost any color, including a wide range of yellows, oranges, reds, browns, and black, as well as greens and blues. Dyeing is rarely identified in practice. The surface concentration of the color may be noted on immersion, although amber naturally darkened by surface oxidation may also show such an effect. Pressed amber usually shows a 'roiled' structure derived from the flow during pressing. There also may be irregular clear and cloudy areas, the normally round gas bubbles may be elongated and/or flattened, and there may even be strain birefringence, seen between crossed polarizers. However, none of these defects may be visible in the best pressed material. Pressed amber usually shows an exceptionally brilliant chalky blue fluorescence. Fluorescence is usually absent with coated amber.

Amber, as well as the more modern resins, gives an aromatic odor when heated, as distinct from the acrid odor of artificial resins and plastics. This can be checked while testing with a hot point (the more modern resins also fuse more readily than amber) or when checking the sectile characteristics with a knife, if there is an inconspicuous part of the specimen available for this test. The more modern resins, such as copal, are more readily attacked by ether than is amber. Careful microscopic examination usually reveals the presence of composite stones, inserted insects, etc., without any difficulty.

References

1. N. Vávra, Bernstein und andere fossile Harze, *Z. dt. Gemmol. Ges,* **31,** 213 (1982)
2. P. C. Rice, *Amber,* Van Nostrand, New York (1980)
3. P. Dahms, Mineralogische Untersuchungen über Bernstein, *Danzig, Schr.,* **8,** 97 (1892–1894); **9,** 1 (1895–1898)
4. G. C. Williamson, *The Book of Amber,* pp. 181–182, 233, Ernest Benn, London (1932)
5. L. Schmid, *Bernstein,* pp. 876–877, 896–899, Steinkopf, Dresden (1931)
6. M. Bauer, *Precious Stones*, p. 539, Dover, New York, Vol. 2 (1968)
7. D. E. Eichholz, *Pliny, Natural History,* Vol 10, Book 37, Chs 11, 12, pp. 199–203, Harvard University Press, Cambridge, MA (1962)
8. R. T. Liddicoat, Jr, Developments, *Gems Gemol.,* **12,** 60 (1966); also **13,** 67 (1967)
9. K. Andrée, Über Inklusen im allgemeinen und über Bernsteininklusenfälschungen im besonderen, *Bernstein-Forsch,* **4,** 52 (1939)
10. W. B. Dick, *Dick's Encyclopedia of Practical Receipts and Processes or How They Did It in the 1870s,* Funk and Wagnall, New York (1974)
11. C. F. Fryer, Amber in plastic, *Gems Gemol.,* **19,** 171 (1983)

Amblygonite

Colorless material has been reported by Pough to turn a greenish yellow color on irradiation[1]. Heat could be expected to reverse this change.

Reference

1. F. H. Pough, The coloration of gemstones by electron bombardment, *Z. dt. Ges. Edelsteink.*, **20,** 71 (1957)

Amethyst, *see* quartz

Andalusite

It is reported that olive-green material from Brazil changes to pinkish when heated[1], while brown material is reported[2] to fade above 800 °C to colorless. Irradiation could conceivably reverse these changes.

References

1. J. Sinkankas, *Gemstone and Mineral Data Book,* p. 113, Winchester Press, New York (1972)
2. G. Smith, E. R. Vance, Z. Hasan, A. Edgar, and W. A. Runciman, A charge transfer mechanism for rose quartz, *Phys. status solidi,* **A46,** K135–K140 (1978)

Anhydrite, *see* celestite

Animal Bones, Teeth, Tusks, *see* ivory

Apatite

Some violet apatite turns green on exposure to radium[1] and violet apatite which has lost its color from heating (perhaps to 500 °C) is turned partly green and partly violet; there is no effect of electrons and only a minor effect of X-rays on some yellow and green specimens[2, 3].

References

1. K. Przibram and J. E. Caffyn, *Irradiation Colours and Luminescence,* p. 244, Pergamon Press, London (1956)
2. F. H. Pough and T. H. Rogers, Experiments in X-ray irradiation of gem stones, *Amer. Min.*, **32,** 31 (1947)
3. F. H. Pough, The coloration of gemstones by electron bombardment, *Z. dt. Ges. Edelsteink.*, **20,** 71 (1957)

Apophyllite

Colorless material has been reported by Pough[1] to turn green on irradiation. Heat could be expected to reverse this change.

Reference

1. F. H. Pough, The coloration of gemstones by electron bombardment, *Z. dt. Ges. Edelsteink.*, **20,** 71 (1957)

Aquamarine, *see* beryl

Aragonite, *see* marble *and* pearl

Aventurine Feldspar, *see* feldspar

Aventurine Quartz, *see* quartz

Barite, *see* celestite

Benitoite

The deep blue color observed in some benitoite may have the same iron–titanium charge-transfer origin as does the blue in sapphire. It is doubtful that any of the corundum heat treatments could be applied successfully, however, since benitoite has a much lower melting point, being a silicate of composition $BaTiSi_3O_9$. G. Rossman has found (unpublished data) that benitoite gradually loses its blue color on heating in air at 600 °C for 19 h. Irradiation with cobalt 60 returned some of the color, but this faded on exposure to light and may be due to a color center that is unrelated to the natural blue color.

Beryl

Including: aquamarine, colorless beryl, emerald, golden beryl, goshenite, heliodor, morganite, red beryl, etc.

Summary of Treatments

Heat:
- **A** Remove yellow: green to blue (aqua); yellow to colorless; orange to pink; all [s, r]
- **B** Remove pink (morganite) [s, r]
- **C** Remove Maxixe blue: blue to pink, etc.; green to yellow; both [s, r]

Irradiation:
- **D** Add yellow color: blue to green (aqua); colorless to yellow; pink to orange; all [s, r]
- **E** Add or increase pink (morganite) [s, r]
- **F** Add Maxixe blue: pale colors to deep blue; yellow to deep green; both [f, r]

Other:
- **G** Synthetic overgrowth [u*]
- **H** Dyeing [s or u, f]
- **J** Colorless impregnation [u]
- **K** Colored impregnation [u, f]
- **L** Composites [s or u, f]
- **M** Coating, foil, etc. [u, f]

*Unstable only in the sense that it may be lost on recutting or repolishing.

The best single reference source on the beryls is the comprehensive text of Sinkankas[1]. The beryls constitute a large family of gemstone materials of the formula $Be_3Al_2Si_6O_{18}.xH_2O$, where x can vary from 0 to 2; some water is almost always present but it is often omitted from the formula. The structure contains channels outlined by rings of $(Si–O)_6$ units; these rings are stacked above each other so that there is room within the channels for water and other impurities[1,2]. Beryl is colorless (called goshenite) when free of the color-causing impurities chromium, vanadium, iron, manganese, or of any irradiation-induced coloration; cesium does not produce any color, as was previously suspected.

Beryl is colored an emerald green by the impurities chromium and/or vanadium; both are frequently present, and there is a lack of agreement as to how emerald should be defined[1,2]. In the opinion of some it is the presence of chromium that is the deciding factor; others use the depth of the color as the criterion, whatever the color-causing impurity. Neither heat nor irradiation affects the chromium- or vanadium-caused colors, although it may be possible to remove a trace of yellow if this is produced by the additional presence of iron, as described below.

Manganese can cause a pink morganite to a deep-red beryl color. Details have been given elsewhere[2,3]. The manganese-produced pink color of morganite fades with heating to about 500 °C[4,5], but even so it can be modified at a lower temperature if some iron-produced yellow is present as well, as described below. **B**

Iron can cause a wide range of colors, including yellows, greens, and blues. As discussed in Chapter 5, p. 57, iron is present within beryl in two types of locations. One type is located on an aluminum site and gives a yellow color if present as Fe^{3+} or absorbs no light and is therefore colorless if present as Fe^{2+}. Heating produces the change from Fe^{3+} to Fe^{2+} and, accordingly, results in the change from the yellow of yellow beryl (also golden beryl or heliodor) to colorless beryl. The other type of iron is situated in a channel site and gives a blue color that is unaffected by heating. If both types are present, heating changes green aquamarine, the color of this gemstone that was once considered most desirable, to the currently preferred blue aquamarine (*see Plate XI*). Heating beryl that contains both iron and manganese analogously changes orange-colored beryl or morganite into the more desirable pink morganite. These heat treatments may be conducted with the materials buried in charcoal or sand, in a glass test-tube, or even by baking the stone while imbedded inside bread dough or clay balls[5]. Some care must be taken in the heating so that cracking does not occur. Temperatures from 250 to 500 °C are suitable, with one to a few hours required at the lower temperatures and just a few minutes at the higher ones[5,6]; temperatures below 400 °C are safest. Some material from India and Brazil is said to require temperatures as high as 700 °C. Reducing conditions are frequently used[2,4], but usually are not necessary. Small inclusions can sometimes be removed by such a heat treatment. **A**

All of these heat-induced changes are stable to light, but can be reversed on irradiation by gamma rays, X-rays, electrons, etc., as was noted by Pough and Rogers[7]. If the necessary iron is present as Fe^{2+} in the substitutional site as found, then irradiation produces yellow from colorless, green from blue (*see Plate XI*), and orange from pink. These colors are all stable to light, although a slight initial fading **D** which does not proceed any further has been reported. Many references to early and sometimes very bewildering studies are given by Sinkankas[1]. Additional references, many with earlier confusing and inconsistent interpretations of the underlying mechanism of the color change, including the unconfirmed suggestion of a yellow-producing color center, have been given by Schmetzer *et al.*[4] Sinkankas[8] mentions peach-colored morganites from Brazil and California which fade in the light to a permanent pink, while Frondel[5] reports that the pink completely disappeared on heating 495 °C. **D**

Care must be taken not to overheat beryl; a milkiness sets in as low as 550 °C if much water is present in the beryl, or there may be no alteration even above 1000 °C in the water-free red beryl from Utah or in synthetic flux-grown material.

One curious report is that by Bank[9] where an Inamori synthetic emerald turned black on irradiation; he also noted a darkening on exposing another synthetic emerald to X-rays which, however, faded after a few hours or days to the original color[9]. No specific cause has been suggested for these changes; one wonders if these stones could have been oiled so that the darkening might have occurred in the oil.

Irradiation restores the pink color of heat-bleached morganite. It is accordingly conceivable that some natural colorless beryl could be turned into morganite or that very pale pink morganite could have its color intensified by irradiation; there do not appear to be any reports of such experiments. **E**

Maxixe and Maxixe-type beryls have a deep blue, almost cobalt blue color, about the equal of the color of sapphire; these colors occur in nature, but can also be produced by irradiation. The color fades when exposed to light. The original natural crystal appears to have been found about 1917, at the Maxixe Mine in Brazil, but gemological examinations were not published until much later[10–12]. Details of this original material have been published recently by the author and co-workers[13–15] and by Schiffman[16]. **F**

About 1971 similar, but not identical, material appeared in the jewelry trade and has been called Maxixe-type beryl[13–15]; it is sometimes sold under the trade name 'Halbanita' and has been stated[17] to be produced by a 'simple process' using pink beryl from a mine at Barra de Salinas, Minas Gerais, Brazil.

Detailed examination has shown that the blue color is caused by a color center which can be produced in material containing a suitable precursor; specimens from various localities in Brazil, North Carolina, and Rhodesia[15] produce this color. The precursor appears to be a nitrate impurity[18] in the original Maxixe material and a carbonate impurity in more recent Maxixe-type material[18, 19]. Both of these ions have four atoms with 24 electrons in their outermost shells and both can lose one electron on irradiation to form a 23-electron hole center:

$$NO_3^- \rightarrow NO_3 + e^- \qquad (7.1)$$

and:

$$CO_3^{2-} \rightarrow CO_3^- + e^- \qquad (7.2)$$

The liberated electron becomes trapped at some electron center, possibly a hydrogen ion, to form a hydrogen atom:

$$H^+ + e^- \rightarrow H \quad (7.3)$$

F

Colorless or pink starting materials give a deep blue while yellowish starting materials give a bluish green or green color, as shown in *Plates III* and *XI.*

This electron center is not very stable, however, so that even light can release the electron again by the reverse of equation (7.3); this then causes the reverse of equations (7.1) or (7.2) to occur with a loss of color. Although there is some difference in the stability of different specimens, all that have ever been reported have faded in light, typically to half the color depth in one week in bright sunlight (*see Plate III*), and much more slowly in a poorly lit environment[20]. Curiously enough, a deep grayish–blue irradiated beryl was examined by Liddicoat in 1968[21]; although it was not stated whether this stone faded, the spectrum given is that of a Maxixe stone.

Many types of irradiation can be used to produce these color centers. As shown by the slight radioactivity in some early commercial samples[14, 15], neutrons from a nuclear reactor have been used; X-rays, gamma rays, and electrons have also proved suitable, and the 'simple process' used in making 'Halbanita' has been shown by the author and others[18] to involve merely exposure, for one to several weeks, to ultraviolet rays. The identification characteristics are discussed below.

C

Not only light, but also heat bleaches the colors of Maxixe and Maxixe-type beryls; irradiation can once again return the color. Detailed absorption spectra have been given[14, 15]; a typical set of stones originally matched in color is shown in *Plate III* where the stone at the center was kept in the dark, the left stone was exposed to bright sunlight for one week, and the stone at the right was placed in boiling water for half-an-hour. In the dark at room temperature this color center seems to be stable almost indefinitely. It is reported (unpublished information from P. C. Keller, G. Rossman, and J. H. Borden) that the pink morganite found in the early 1970s in Itatiaia, Brazil, is a gray to bluish color as found, but becomes pink on exposure to the sun. A check of the spectrum (D. L. Wood and K. Nassau, unpublished information) on a specimen obtained from J. H. Borden demonstrated that this is indeed a Maxixe beryl, containing nitrate.

G

The growth of a layer of emerald over the surface of a pale-colored beryl, aquamarine gemstone or crystal has been described in full detail in the author's *Gems Made by Man*[22]. Both hydrothermal and flux growth have been used, as briefly described above in Chapter 5, pp. 77–78, where further references may be found to the Lechleitner and Nacken techniques. The product is stable.

H

It is possible to have dyed beryl and emerald where the color lies not in the oil, but is deposited in cracks by the evaporation of a solvent or by chemical reaction; this technique may have been used but has not usually been distinguished from the colored impregnations discussed below.

J

The impregnation of emerald with a colorless oil (or wax or plastic) to make fractures less conspicuous or to completely hide them has been practiced since antiquity, as discussed in Chapter 2. The techniques of cleaning followed by impregnations, often with the use of vacuum, are described in Chapter 5, pp. 61–65. A particularly good detailed account of the technique as used in Colombia has been given by Ringsrud[23]. He reports that the gems are first cleaned by boiling several times with methyl or ethyl alcohol. Next aqua regia is applied, sometimes in a stoppered container (a very foolish practice – *see* footnote on p. 62) and/or using ultrasonics. After a thorough rinsing and drying, one of a variety of oils is applied, often with the use of vacuum, pressure, and/or heat. These oils include '3-in-1' lubricating oil, clove oil, mineral oil, cedarwood oil, and Canada balsam; the optical properties of these oils are listed and discussed in Chapter 5, pp. 64–65. A final polish, sometimes using vaseline, completes the process. Ringsrud also discussed the fact that, according to the dealers, most stones are only slightly enhanced in value and that very few of these stones dry out rapidly, predominantly those with large cracks. A similar, but less detailed, account of the same operation has appeared previously[24]. Wax or plastics could also be used as discussed in Chapter 5. The appearance of oil in cracks is shown in *Figure 7.1*. **J**

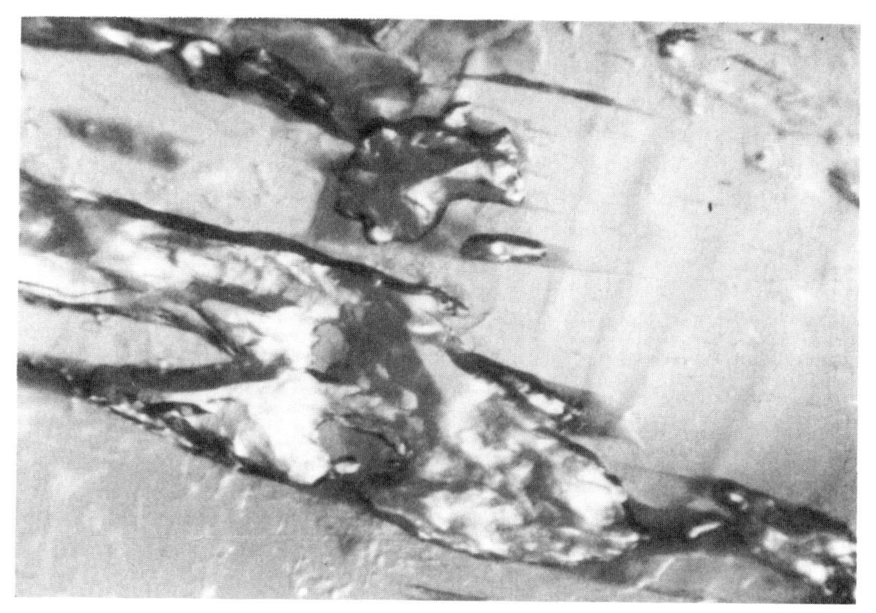

Figure 7.1 Oil filling the cracks of an emerald; it is used to make the cracks less prominent. Photograph by courtesy of the Gemological Institute of America

Less frequent, but more dangerous to the unwary, is the use of coloured impregnations, where a green dye has been added to the oil (or wax, or plastic). This process goes back at least to Roman times; some details of early techniques are given in Chapter 2, pp. 7, 10–12, as well as in Chapter 5, p. 65. One example where both green and yellow dyes were used has been described recently with excellent illustrations[25], which are reproduced in *Plate XXVII*; although usually only an improvement in color is sought, in this instance a colorless, faceted beryl was 'improved' to become an 'emerald'. Also seen occasionally are cabochons of pale-green massive beryl, but sometimes even of colorless (white) massive beryl which have some porosity, that have been impregnated with colored oil[25]. **K**

L

A variety of composite stones are made using beryl. There are doublets of emerald combined with pale beryl or other natural or synthetic substances. A triplet may be made of two parts of beryl cemented together with a dark green dye layer; the inclusions seen in such a stone are, of course, those of beryl. They can be particularly misleading when made of a faceted beryl that has been cut in half and reassembled with the colored cement so that there are no discontinuities in the distribution pattern of the inclusions[26]. General details on composite stones are given in Chapter 5, pp. 75–77. A large 'emerald' crystal has been reported which was made by drilling out a near-colorless beryl and filling it with green cement and crystalline material[27].

M

Colorless and green reflective foils and other backings are only rarely seen in beryls, but surface coatings appear occasionally, usually in a colorless or pale beryl covered with a green plastic coating[28].

#

Identification begins with conventional gemological testing to establish the presence of beryl and, for emerald, to rule out the possibility of synthetic emerald. The detection of the use of heat or irradiation to produce the change among the pairs of colors green–blue aquamarine, yellow–colorless beryl, orange–pink morganite, or colorless–pink morganite is not feasible in practice, there being no consistent indications that do not apply equally to at least some naturally occurring material.

Synthetic overgrowth is readily identified by the change in the inclusion pattern on focusing down into the stone, by the concentration of color in the surface, and by the many fine cracks always present in the Lechleitner product. The occurrence of inclusions typical of beryl rather than of emerald in a dark green stone also leads to the suspicion that an overgrown, coated, or composite gemstone may be involved. The further confirmation of these possibilities follows conventional gemological practice.

Colored oiling can usually be recognized by the localization of color on immersion in water or other fluids or on viewing by the light transmitted through a white, translucent plastic diffuser[25]. The spectrum of the dye rather than the typical chromium spectrum is also an identifying characteristic. Colorless oiling, as in *Figure 7.1*, may present some identification difficulties, particularly if the oil does not fluoresce. In the absence of a perfect refractive-index match, a very careful microscopic examination often reveals a faint outline of the filled cracks. An oily odor may be present, particularly when opening a parcel, and seepage may be seen on carefully holding a hot point near the suspected region of oiled cracks. The emerald filter may give clues in the hands of an expert, but its indication cannot always be relied upon. A solvent on a swab usually removes some of the oil and thus reveals the crack in a 'slightly destructive' test.

Maxixe and Maxixe-type deep blue beryl is readily identified by the spectroscope and the pleochroism. The deep blue color can be taken as a clue, but this should not be relied upon, since a partially faded stone may resemble an aquamarine quite closely. In the spectroscope these materials show quite sharp absorption lines, a strong one between 685 and 695 nm and several weaker lines extending from 670 to below 585 nm; their visibility depends on the nature of the material and the intensity of the coloration. By rotating the specimen in a dichroscope, in a polariscope with the polarizers in parallel position, or merely by using a piece of sheet polarizer, as in *Figure 7.2*, the pleochroism can be determined. When rotating the specimen about the line of view while looking down the optic axis, as in *Figure 7.2(a)*, there is no change in the color when only the ordinary ray is being viewed. This is the lighter of the two dichroic colors in aquamarine, but the darker of the two in Maxixe and Maxixe-type beryl. When

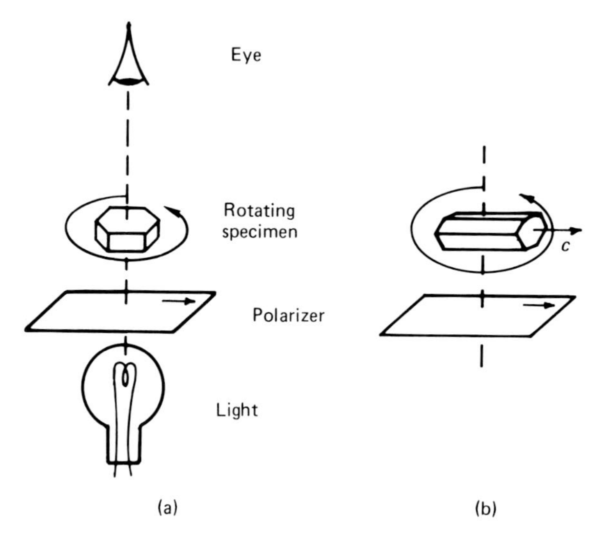

Figure 7.2 Distinguishing between aquamarine and the Maxixe beryls by observing the dichroism while rotating a specimen above a polarizer. (a) Rotation about the optic axis and (b) perpendicular to it; in the position shown at (b), aquamarine shows its darker color, while the Maxixe beryls show their lighter color

viewed with the optic axis at right angles to the sighting direction, rotation about the line of view produces both colors in both of these materials. The color seen when the optic axis is parallel to the axis of the polarizer, as in *Figure 7.2(b)*, is due to the extraordinary ray; this is the darker blue with aquamarine and the lighter with the Maxixes. Immersion of the specimen in water or a refractive-index matching fluid may help with this examination. Heating the specimen in water to the boiling point for less than an hour fades the Maxixes, but this cannot be recommended as a routine gemological test!

References

1. J. Sinkankas, *Emerald and Other Berlys,* Chilton, Radnor, PA (1981)
2. D. L. Wood and K. Nassau, The characterization of beryl and emerald by visible and infrared absorption spectroscopy, *Amer. Min.,* **53,** 772 (1968)
3. K. Nassau and D. L. Wood, An examination of red beryl from Utah, *Amer. Min.,* **53,** 801 (1968)
4. K. Schmetzer, W. Beredinski, and H. Bank, Über die Mineralart Beryll, ihre Farben and Absorptionspektren, *Z. dt. Gemmol. Ges.,* **23,** 5 (1974)
5. C. Frondel, Effect of heat on the colour of beryl, *Gemmologist,* **21,** 197 (1952)
6. G. O. Wild, The treatment of gem stones by heat, *Rocks and Miner.,* **7,** 9 (1932)
7. F. Pough and T. H. Rogers, Experiments in X-ray irradiation of gem stones, *Amer. Min.,* **32,** 31 (1947)

8. J. Sinkankas, Color changes in gemstones, Part 4, *Lap. J.*, **17,** 616 (1963)
9. H. Bank, Aus der Untersuchungspraxis, *Z. dt. Gemmol. Ges.*, **31,** 97 (1982)
10. G. O. Wild, Mitteilung über ein anscheinend neues Berylliumsilikat, *Z. Miner. Geol. Palaönt.*, **1933A,** 38 (1933)
11. K. Schlossmacher and H. Klang, Der Maxixeberyll I, *Z. Miner. Geol. Paläont.*, **1935A,** 37 (1935)
12. W. Roebling and H. W. Tromnau, Maxixeberyll II, *Z. Miner. Geol. Paläont.*, **1935A,,** 134 (1935)
13. K. Nassau and D. L. Wood, Examination of Maxixe-type blue and green beryl, *Gems Gemol.*, **14,** 130 (1973)
14. K. Nassau and D. L. Wood, The nature of the new Maxixe-type beryl, *Lap. J.*, **27,** 1032 (1973)
15. K. Nassau, B. E. Prescott and D. L. Wood, The deep blue Maxixe-type color center in beryl, *Amer. Min.*, **61,** 100 (1976)
16. C. A. Schiffman, Gemmologische Studie an einem Original-Maxixe-Beryll, *Z. dt. Gemmol. Ges.*, **26,** 135 (1977)
17. F. M. Bastos, Maxixe-type beryl, *Lap. J.*, **28,** 1540 (1975)
18. L. O. Anderson, The difference between Maxixe beryl and Maxixe-type beryl: An electron paramagnetic resonance investigation, *J. Gemm.*, **16,** 313 (1979)
19. A. Edgar and E. R. Vance, Electron paramagnetic resonance, optical absorption, and magnetic circular dichroism studies of the CO^- molecular-ion in irradiated natural beryl, *Phys. Chem. Miner.*, **1,** 165 (1977)
20. K. Nassau and B. E. Prescott, Nonfading Maxixe-type beryl?, *Gems Gemol.*, **17,** 217 (1981)
21. R. T. Liddicoat, Jr, Irradiated spodumene and morganite, *Gems Gemol.*, **12,** 315 (1968)
22. K. Nassau, *Gems Made by Man,* pp. 131–140, 149–150, Chilton, Radnor, PA (1980)
23. R. Ringsrud, Emerald treatment in Bogota, Colombia, *Gems Gemol.*, **19,** 149 (1983)
24. Anon, Secrets of emerald oiling, *Retail Jeweller,* **20,** 2 (Apr 9, 1981)
25. C. F. Fryer (Ed.), Emerald: emerald substitute, dyed beryl, *Gems Gemol.*, **17,** 227 (1981)
26. R. Crowningshield, Beryl-and-beryl triplet, *Gems Gemol.*, **10,** 69 (1960)
27. C. F. Fryer (Ed.), Fake specimen, *Gems Gemol.*, **18,** 44 (1982)
28. R. T. Liddicoat, Jr, Coated beryl, *Gems Gemol.*, **13,** 320 (1971)

Bloodstone, *see* chalcedony

Blue John, *see* fluorite

Bone, *see* ivory

Bone, Petrified, *see* chalcedony

Bowenite, *see* marble

Brazilianite

This material has been reported to fade[1] and to lose its greenish color completely on being heated above 140 °C[2]; irradiation did not reverse this change[2].

References

1. J. Sinkankas, *Gemstone and Mineral Data Book,* p. 114, Winchester Press, New York (1972)
2. A. Requardt, F. Hill, and G. Lehmann, A firmly localized hole center in the mineral Brazilianite, *Z. Naturforsch.*, **A37,** 280 (1982)

Buergerite, *see* tourmaline

Burinut, *see* ivory

Calcite, *see* marble

Cancrinite

Colorless material has been reported to acquire blue veins on being irradiated with X-rays[1]. One assumes that this is the fibrous variety described by Webster. One would expect this change to be reversed by heat, and fading in light is also a possibility.

Reference

1. F. H. Pough and T. H. Rogers, Experiments in X-ray irradiation of gem stones, *Amer. Min.*, **32**, 31 (1947)

Carnelian, *see* chalcedony

Cat's-eye, *see individual gemstone variety involved*

Celestine, *see* celestite

Celestite

Celestite, $SrSO_4$, also called celestine, as well as the closely related materials anhydrite, $CaSO_4$, also called vulpinite (Webster, p. 298), and barite, $BaSO_4$, are soft sulfates which occur at times with attractive violet or blue colors. Heating to as little as 200 °C causes these colors to fade, as also sometimes happens on exposure to light. Irradiation restores or produces the blue or violet color, sometimes in a banded form; the color may be stable to light or may fade, depending on the nature of the material[1,2]. In celestite there are several color centers involved, including SO_3^-, SO_2^-, and O^-, all stabilized by a potassium impurity[3].

References

1. K. Przibram and J. E. Caffyn, *Irradiation Colours and Luminescence,* pp. 241–242, Pergamon Press, London (1956)
2. F. H. Pough, The coloration of gemstones by electron bombardment, *Z. dt. Ges. Edelsteink.*, **20,** 71 (1957)
3. L. R. Bernstein, Coloring mechanism in celestite, *Amer. Min.*, **64,** 160 (1979)

Ceruleite

This hydrated copper aluminum arsenate[1] resembles turquoise in being frequently porous. Plastic impregnation has been reported[2] as a 'stabilizing' process (*see also* turquoise).

References

1. K. Schmetzer, H. Bank, W. Beredinski, and E. Krouzek, Ceruleite – a new gemstone, *J. Gemm.*, **16,** 86 (1978)
2. K. Schmetzer, T. Lind, and H. Bank, Stabilized ceruleite, *J. Gemm.*, **18,** 734 (1983)

Chalcedony

Including: other cryptocrystalline varieties of quartz such as agate, agatized bone, coral, or wood, bloodstone, carnelian, chert, chrysoprase, flint, heliotrope, jasper, novaculite, onyx, petrified bone, coral, or wood, plasma, prase, sard, sardonyx, etc. (but generally excluding the crystalline quartz family which is covered under quartz).

Summary of Treatments

Heat:
- **A** Pale colors to brown and red [s]
- **B** Pale colors to milky white [s]

Irradiation:
- **C** Darken [s, r]

Other:
- **D** Bleaching [s]
- **E** Colored impregnation [u, f]
- **F** Dyeing [s or u, f]
- **G** Coating, foil, etc. [s or u, f]

Some of these treatments can also be applied to some of the porous crystalline quartzes such as aventurine quartz, cat's-eye quartz, quartzite, tiger-eye quartz, etc., described under quartz. Analogously, some of the treatments there described also apply to some of the cryptocrystalline quartzes discussed here.

Chalcedony and the related materials listed above the summary table are all quartzes, SiO_2, that have a cryptocrystalline (microcrystalline) structure composed of tiny fibrous crystals in a partly amorphous matrix. There is a significant porosity associated with this construction. In the additional list below the summary table are given those crystalline quartzes that may have some porosity so that several of these treatments can also be applied; other treatments listed under quartz should be consulted for these as well. In the account that follows, 'chalcedony' is used as a collective group name. The question of the confused terminology of these materials is considered here as being of no importance to the subject matter; in the words of Anderson[1]: '. . . we have a host of names to contend with for what is essentially the same material'.

A Heat treatment is used on yellow-to-pale-to-brown iron-containing chalcedony, particularly carnelian (also spelled cornelian) to produce rich brown to red colors. This change involves a hydration alteration, usually from limonite to hematite, as discussed in Chapter 3, pp. 32–33.

B Heating can turn bluish or grayish tints as well as other colors to a milky white and this step is used as a preliminary preparation for dyeing. Heating is also a part of several of the dye decomposition processes discussed in Chapter 5, pp. 66–70 and 73. All these changes are permanent and cannot be reversed. Chrysoprase, a chalcedony colored green by the presence of nickel, loses its color on being heated; soaking in water restores the color[2]. **B**

Irradiation has various effects on quartz as discussed under that material; this process does not appear to have been reported for chalcedony, but could be used to modify the color. **C**

If chalcedony contains iron stains accessible to chemicals, bleaching of the color can be achieved by the use of concentrated hydrochloric acid, a saturated water solution of oxalic acid, or some of the other agents listed in Chapter 5, pp. 62–63. **D**

Colorless oiling is rarely seen in chalcedony, but colored impregnations have occasionally been used, employing the techniques described in Chapter 5. **E**

Dyeing is very widely used on the cryptocrystalline quartzes to produce commercial products such as black onyx (rare in nature), Swiss lapis (blue, of a shade never found in nature), 'chrysoprase' and 'carnelian' (quite different from the natural green and red materials of those names), and so on. Colors are produced that imitate almost every opaque or translucent gemstone. Those materials that absorb dye unevenly to produce banded layers of red and white or black and white when dyed are widely used as 'onyx' and 'sardonyx', respectively, for carved cameos and intaglios. **F**

A detailed description of some of the Idar Oberstein dyeing processes has been given in Chapter 5, pp. 66–70. In addition to the references given there, and particularly the extended series of Frazier, there are many other accounts, some of which give practical advice for the amateur dyer, including those by Sinkankas[3], Jones[4], Leiper[5], and Maynard[6]. Some of these, particularly the last, use organic dyes, where familiar names such as rhodamine-B and malachite green[7] are found; such colorations can be expected to fade, while the precipitated or reacted inorganic colorants used at Idar Oberstein are quite stable to light, as described in Chapter 5.

Dyeing has also been used to produce specific patterns. Stones which closely resemble naturally included chalcedony can be produced by painting dyes onto a flat or curved chalcedony surface. Close inspection, however, reveals the process. A silhouetted head[8] and dendritic patterns have been reported[8, 9]; silver staining, as described in Chapter 5, was apparently used on both occasions, with a possible pretreatment with a salt (sodium chloride) solution, as described by Michel[10].

Apart from varied chalcedony colors such as onyx, carnelian, and so on, dyed quartzes are also used to imitate jade, lapis lazuli, and other materials. Unnatural colors are one of the ways of recognizing dyed material and the gemologist can frequently recognize the dye used by eye or by the spectroscope; the employment of a new dye is often reported in the gemological literature when it is first seen[11, 12].

The once widely used foil-backs are only rarely seen today, most often in a translucent stone with an engraved back to simulate a star corundum. Coated stones, particularly the carnelians decorated with white patterns produced by the use of sodium carbonate and heat, are described in Chapter 5, p. 73. S. Frazier (unpublished observation) reports an agate cabochon with dendrites engraved into the back, which was then filled with a black substance. **G**

Chalcedony that has been dyed or otherwise treated may substitute for the untreated material or for a wide variety of other minerals, including the amazonite and moonstone feldspars, jade, lapis lazuli, malachite, and turquoise. In themselves, the rather intense, at times garish, dyed colors are usually quite obvious to the experienced gemologist's eye. Also the even coloration of, say, green, dyed chalcedony compared to the slightly mottled appearance of chrysoprase quartz can be a definite clue; the former may sometimes also have a pale pink color when viewed through the emerald filter. Weak chromium lines may be seen in the spectroscope of any chromium-dyed material. Essentially, all 'black onyx' is sugar-dyed chalcedony. Distinction from other materials follows conventional gemological testing, with the refractive index and the specific gravity being important parameters. Heating and dyeing of chalcedony are so widely accepted in the trade that the attempt to identify these treatments is rarely made.

References

1. B. W. Anderson, *Gem Testing*, pp. 316–317, 9th Edn, Butterworths (1980)
2. M. Bauer, *Precious Stones*, Vol 2, p. 497, Dover Publications, New York (1968)
3. J. Sinkankas, Color changes in gemstones, Part 3, *Lap. J.*, **17**, 532 (1963)
4. B. Jones, Dyeing agates, *Rocks Gems*, **7**, 60 (1977)
5. H. Leiper (Ed.), *The Agates of North America*, Lapidary Journal Press, San Diego, CA (1966)
6. N. Maynard, New methods of coloring agates and cabochons, *Lap. J.*, **12**, 640 (1958)
7. K. Nassau, *The Physics and Chemistry of Color: The Fifteen Causes of Color*, Wiley, New York (1983)
8. P. C. Zwaan, An unusual dyed agate, *J. Gemm.*, **9**, 283 (1965)
9. R. T. Liddicoat, Mossifying chalcedony, *Gems Gemol.*, **12**, 118 (1966–1967)
10. H. Michel, *Die Künstlichen Edelsteine*, pp. 378–380, W. Diebner, Leipzig (1926)
11. R. T. Liddicoat, A new color in dyed chalcedony, *Gems Gemol.*, **11**, 372 (1965–1966)
12. R. Crowningshield, Dyed chalcedony carving, *Gems Gemol.*, **13**, 350 (1971)

Chert, *see* chalcedony

Chloromelanite, *see* jade

Chrysocolla

Including: Eilat or Elath stone.

These copper-containing minerals have blue-to-green colors and occur together with other copper minerals, as well as in a disseminated form in chalcedony, opal, and quartz (Webster). Dyeing has been used to enhance the color and to make it resemble turquoise or other gemstones.

Chrysoprase, *see* chalcedony *and* quartz

Citrine, *see* quartz

Coco de Mer, *see* ivory

Conch Pearl, *see* pearl

Copal, *see* amber

Coral

Summary of Treatments

A Dyeing [s or u, f]
B Bleaching from black to gold [?]
C Coating or filling [s or u]
D Molding [s]

The classic white-to-red colored coral, *Corallium rubrum* or *nobile* is the branched framework of the marine coral polyp. It is calcareous, that is composed of calcite, $CaCO_3$, with small amounts of magnesium carbonate, $MgCO_3$, and traces of iron, assembled in a characteristic radiating structure; the pigment is provided by 1.5 to 4 percent organic matter, conchiolin, a substance similar to the conchiolin in pearls.

In addition, there are other white-to-pink corals such as *Corallium secundus* and the species *Oculinacea*. Lastly, there are the blue coral *Allopora*, the gold-colored species *Gerardia* and *Porazoanthus* (dimpled surfaces), and several dark to black corals, including *Antipathes* and *Cirrhipathes* (spiny surfaces). These variously colored blue, golden, and black species are horny in nature, being composed predominantly of conchiolin, again with the color derived from this organic substance. Reviews are found in Webster and Bauer[1]; also relevant are articles by Brown[2,3] and Aliprandi *et al.*[4].

A Dyeing has been used to produce the desired 'ox-blood' (dark red) and 'angel skin' (pink without any trace of orange) colors from colorless or pale colored *Corallium* and *Oculinacea* coral. The color may be applied only in selected areas[5] and may be stable or may fade.

B Bleaching with 30 percent hydrogen peroxide solution for 12 to 72 h[3] is used to convert the black varieties of coral to a very attractive golden color. The stability of this bleached color with time and exposure to light does not appear to have been studied.

C Epoxy-type resins have been used to fill the cavities of poor-quality coral[4], thus permitting the shaping and polishing of otherwise unsatisfactory material. A colored coating has also been used to cover white spots[6].

D Bauer[1] reports, without detail, that large pieces of black coral, being of a horny nature, can be '. . . moulded to form armlets and the like'.

Identification of coral is simplified by its characteristic grained, radiating or tree-ring structures. The specific gravity of the calcareous types is 2.60 to 2.70, while that of the horny types is much lower. The very soft black and golden types emit the odor of burnt hair when tested with the hot point. In the unpolished material, the surface characteristics can help to distinguish the natural gold (dimpled surface) from the bleached black (smooth or spiny)[3]. If the bleach has not penetrated fully, a gold–black boundary may be visible. The bleached material tends to have lower values of the specific gravity (1.35 to 1.40 *vs.* 1.45) and the refractive index (1.55 to 1.56 *vs.* 1.56) than does the natural, gold-colored material[3]. In cross-section, the natural gold shows dashed radial features, while the # bleached black has tree-ring or radial spinous structures[3]. The presence of coatings or fillings can be detected by careful gemological testing; the specific gravity is usually lowered by added material. Dye has been detected with nailpolish remover[5], ether, or acetone[4]. A weak, purplish red fluorescence under long-wave ultraviolet has been noted[4]. There are many imitations, including one that is not too far from a 'synthetic'[7].

References

1. M. Bauer, *Precious Stones,* Vol 2, pp. 601–615, Dover Publications, New York (1968)
2. G. Brown, Gold corals – some thoughts on their discrimination, *Gems Gemol.,* **16,** 240 (1979–1980)
3. G. Brown, Golden corals: a brief note, *Austral. Gemm.,* **14,** 204 (1981)
4. R. Aliprandi, F. Burragato, and G. Guidi, Natural coral and some substitutes, *J. Gemm.,* **18,** 401 (1983)
5. R. Crowningshield, Dyed angel's-skin coral, *Gems Gemol.,* **12,** 209 (1967)
6. R. T. Liddicoat, *Handbook of Gem Identification,* p. 393, 11th Edn, Gemological Institute of America, Santa Monica, CA (1981)
7. K. Nassau, An examination of the new Gilson 'coral', *Gems Gemol.,* **16,** 179 (1979)

Coral, Petrified, *see* chalcedony

Cordierite, *see* iolite

Cornelian, *see* chalcedony

Corozo Nut, *see* ivory

Corundum

Including: ruby, variously colored sapphires, including padparadscha and asteriated varieties.

Summary of Treatments

Heat:

- **A** Develop or intensify iron-caused yellow; turn pink to orange, blue to green; all [s]
- **B** Lighten or remove color-center caused yellow; turn orange to pink, green to blue; all [s, r]
- **C** Silky and asteriated to clear [s, r]
- **D** Silky and clear to asteriated [s, r]
- **E** Develop or intensify a blue component in sapphire and ruby [s, r]
- **F** Lighten or remove blue component in sapphire or ruby [s, r]
- **G** Even out Verneuil banding [s]
- **H** Introduce fingerprint inclusions [s]

(continued)

(continued)

Heat plus additive:

J In-diffusion of colors and/or asterism [u*]

Irradiation:

K Colorless to yellow, pink to orange, blue to green; (YFCC); all [f, r]

L Colorless to yellow, pink to orange, blue to green; (YSCC); all [s, r]

Other:

M Synthetic overgrowth [u*]

N Dyeing [s or u, f]

O Colorless impregnation [u]

P Colored impregnation [u, f]

Q Composites [s or u, f]

R Coating, foil, applied or inscribed 'asterism', etc. [s or u]

*Unstable only in the sense that it may be lost on recutting or repolishing.

There is no comprehensive, single treatise dealing with the corundum family of gemstones. Texts such as Webster's *Gems* and the author's *Gems Made by Man*[1] summarize the occurrence and gemology, and the synthesis and structure, respectively.

Corundum is single-crystal Al_2O_3, aluminum oxide or alumina, which is colorless when free of color-producing transition metals or color centers. In gemstone form it is called sapphire, with a color designation added if any color other than blue. The term ruby is restricted to a limited range of medium to deep reds; the only other term normally used is padparadscha for a pink–orange color[2].

Chromium by itself gives corundum a pink to deep-red ruby color. Iron, when present by itself, can give to corundum a pale green, yellow, or brownish color, but in the presence of titanium also produces a wide range of greens, blue–greens, and blues that originate in a charge-transfer mechanism involving both ions[3]. Many reports[4–8] have summarized the frequently confusing studies published over the years; these have not always shed light on the subject. A brief account is given in Chapter 3, pp. 33–40, where references are listed.

Titanium itself can lead to asterism and to silk; some silk-like inclusions can, possibly, be produced by other impurities as well. Other colors can be produced by other transition-metal impurities, such as vanadium, nickel, and cobalt[1], but these are not normally involved in gemstone enhancement.

There appear to be at least two color centers producing a yellow color, one of which fades in light while the other does not. The designations YFCC for 'yellow fading color center' and YSCC for 'yellow stable color center' are used herein for these color centers to prevent confusion.

Changes Involving Yellow–Brown and Orange–Pink Corundum

A When pale green, pale yellow, or near colorless corundum containing some iron is heated, then a yellow-to-gold-to-brown color (hereafter merely termed yellow) is produced, as indicated in *Figure 7.3(a)*. This change is possible even in the presence of titanium if the corundum is fully oxidized or if the iron content is much in excess of the titanium content. It is clear that this color originates from Fe^{3+}, although this is not necessarily substitutional iron and may not be required in very large amounts. This yellow color is frequently much darker and richer than that of the usual, naturally occurring yellow sapphire; it is stable to light and heating, and probably cannot be reversed once formed.

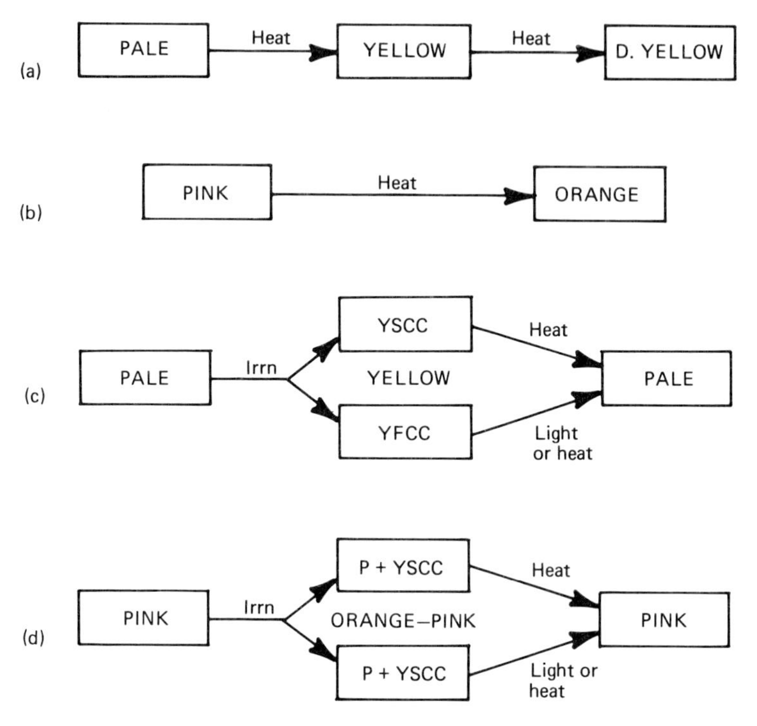

Figure 7.3 The heat and irradiation changes involving yellow colors in sapphire; YSCC stands for yellow stable color center and YFCC for yellow fading color center: (a) iron present alone; (b) iron plus chromium; (c) color center alone; (d) chromium plus color center

Temperatures required for this conversion range from below 1500 °C, where an extended period is required, to about 1900 °C, where only a few minutes suffice, but where there is a great danger of melting the corundum in the absence of good temperature control. Heating is normally performed in an open crucible in air to oxidize Fe^{2+} to the desired Fe^{3+}. Chemicals do not need to be used, but even so damage to the surface occurs at such high temperatures. Some details have been presented in Chapter 3, pp. 33–35. A description of the practice in Thailand has been given by Keller[9] and additional data appears elsewhere[7, 8]. A heat-treated yellow sapphire said to fade[10] seems highly improbable; it is much more likely that this material was in fact irradiated. With a pink corundum colored by chromium with some iron also present, this heat treatment leads to similarly stable padparadscha orange–pink colors, as in *Figure 7.3(b)*.

K,L Irradiation by almost any of the techniques described in Chapter 4 and Appendix B produces color centers of two types, one fading, the other stable. Almost nothing is known about the chemical or physical nature of these color centers. Both can occur simultaneously, as in the yellow sapphire[11] which lost its color on being heated during jewelry repairing; radiation on the advice of the author produced a much darker yellow, which faded back to the original yellow on exposure to light, as described in Chapter 4 under The Restoration of Color, pp. 56–57. The fading yellow corundums have been seen occasionally[12–15], although a number of earlier experiments have also been reported, such as that by Pough and Rogers[16]. In *Figure 7.3(c)*, these two color centers are designated YSCC and YFCC for 'yellow stable/fading color center'.

If these yellow colors appear in the presence of some chromium (which causes pink by itself), the orange–pink color of padparadscha appears, as in *Figure 7.3(d)*, and a blue sapphire can similarly be turned green. Some naturally occurring yellow-to-orange sapphires contain the Fe^{3+} yellow, others the yellow stable color center YSCC; since the YFCC fades so rapidly in light, a matter of just a few hours in sunlight, it probably does not occur in the gemstone and jewelry trades in the sense of not surviving a collecting trip unless 'restored' by irradiation. Synthetic corundum sometimes produces a yellow or a brown color on irradiation, but not always; Lehmann and Harder[4] report that some such material fades after a few days, even in the dark.

B Heat removes both color centers, whether in the natural or in the irradiated material. About 500 °C is sufficient[8] to bleach the YSCC and much less is required for the YFCC, which bleaches at 200 °C or less, and in just a few hours' exposure to light. This heating converts yellow to lighter or colorless, padparadscha to pink, and green to blue, if the yellow component in these originated from a color center. Irradiation from almost any source will again restore the original colors.

The Silk–Asterism Changes

C,D Titanium by itself, present in the form of titanium oxide, may enter corundum to produce quite colorless and transparent material, merely silky material, or a 'white' star. The difference between these lies in the limited solubility of titanium oxide in corundum at low temperatures and the good solubility at high temperatures, particularly if the conditions are reducing. A detailed discussion with references is given in Chapter 3, pp. 38–40.

In summary, heating for an extended period at between 1600 °C and 1900 °C makes the star-causing and silk-causing needles dissolve in the alumina and produces clear material; reasonably rapid cooling (a few hours at the most) preserves this condition. The coarser the needles, the longer the heating time that is required for dissolution. Heating for an extended period at about 1300 °C causes these needles to form (or re-form) and thus produces asterism; the longer the heating, the more the needles grow in thickness up to a limit controlled by the titanium concentration and other factors. These effects can, of course, occur in the presence of chromium oxide in ruby, iron oxide in blue- and green-sapphires, and so on. These changes are summarized in *Figure 7.4(b)*. The needle-forming step is an essential part of the manufacture of synthetic star corundum[1].

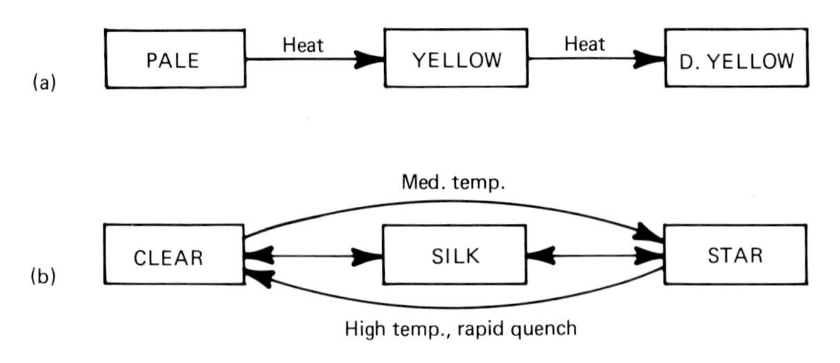

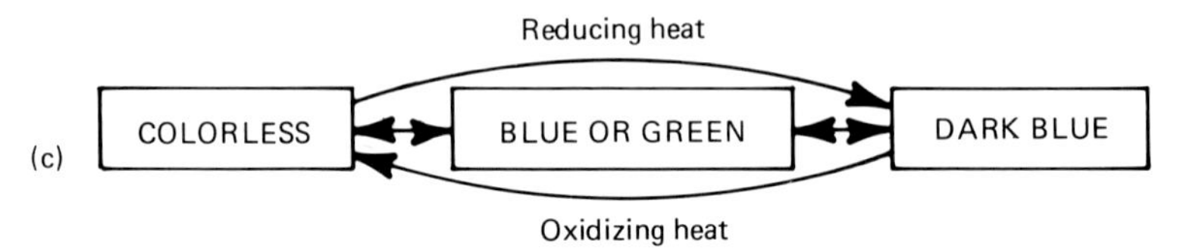

Figure 7.4 The heat and irradiation changes involving iron and titanium in yellow-to-green-to-blue-to-black sapphires, including asterism: (a) iron present alone; (b) titanium present alone; (c) iron plus titanium

The Blue and Green Corundum Changes

E,F

As described in Chapter 3, pp. 33–40, the combination of iron plus titanium oxides in corundum can produce a wide range of colors, varying from colorless via yellow and green to blue–green, blue, and blue–black. As there discussed, the color produced by specific amounts of these impurities is not fixed, but depends on the heating history of the material; this variability depends on many factors, including the maximum temperature reached, the oxidizing or reducing conditions during the heating, and so on. Smaller amounts of other impurities may modify these colors. Smaller amounts of these colors can also be present in ruby, giving it a brown or a purple shade, and so on.

The question frequently asked is: 'Where can I purchase a furnace to perform sapphire heat treatments?' The listing of suppliers given in Appendix D is intended to remove this pressure from the author. At the same time, such a questioner must realize that the furnace is merely the beginning. As described in Chapter 3, placing two equally colored pale-blue sapphires into such a furnace and choosing the atmosphere and temperature at random may well result in one of the two stones reaching a dark blue, with the other turning almost colorless, both under the same conditions! Also the fuels and gases used may be highly explosive if misused or if leaks develop; these hazards cannot be overemphasized.

The remarks that follow are no more than rough guides gleaned from available information and from known scientific principles. Regrettably, almost everyone who has had some success with these processes on the specific rough, views such data as trade secrets, thus severely limiting the technical information available. Despite much variability, the author's outline of heat treatments presented above and based on previously published work[7] is probably the best comprehensive guide available; a diagram which may be helpful was published previously[17] and is reproduced in *Figure 7.5*.

Consider first typical 'Geuda' material[18], whitish or pale blue corundum of poor color and possibly showing some silk. Heating such material in air up to perhaps 1200 °C merely removes what blue there is present. This process was reported in detail in 1658, when Porta described how 'to turn a sapphire into a diamond', and also how to produce a blue–white two-color stone by selective limitation of air access, as described in Chapter 2, pp. 17–18. Recent descriptions include those by Jobbins[19] and Tombs[20]. The process that develops a deep blue color and at the same time removes silk involves heating to a much higher temperature; 1200–1700 °C has been suggested by Tombs[21]; the upper range is

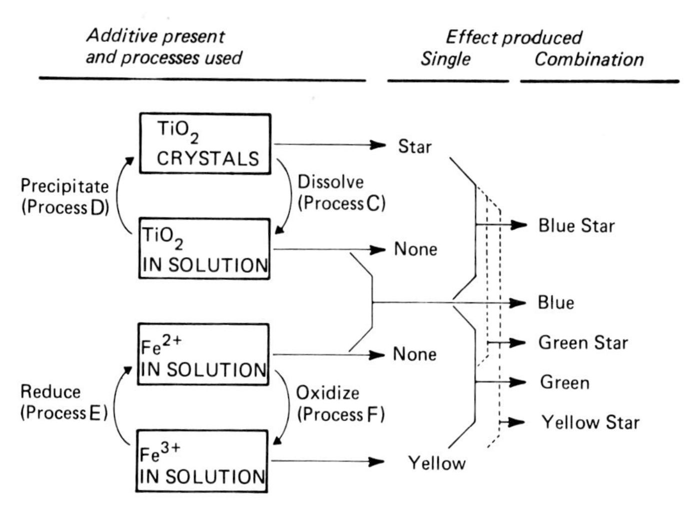

Figure 7.5 Another way of representing the data of *Figure 7.4*

consistent with the 1600 °C given by Abraham[22], presumably involving very long heating times. In view of the occasional accidental melting reported, temperatures may well range up to 1900 °C and even higher for shorter periods, which are economically more favorable in view of the high cost of fuel. A heat-treated specimen, originally almost colorless, is shown in *Plate XIII*.

There appears to be general agreement that the stones are first trimmed or preformed to remove all but the smallest of inclusions and are sometimes soaked, covered, or painted with a borax-based solution. They are next placed into an alumina crucible, often doubled as in *Figure 3.2*, and then surrounded by charcoal (or otherwise arranged, e.g. by using hydrogen-containing gases) to produce a reducing environment according to Abraham[22] and P. C. Keller (unpublished information). Chemicals said to be used in Thailand and Burma as reported to Keller include borax used by itself or in combination with, or replaced by, sodium acetate, sodium or potassium hydroxide, glycerin, citric acid, other acids, potassium nitrate, a chlorine-containing compound, and so on. These additives are said to reduce cracking and even to help heal existing cracks and to expedite the color change. Borax is the most consistently reported substance, although some operators are said to use no chemical at all[22]. Heating may extend from a few hours to a few days. **E,F**

Based on the result obtained in the first heating, the decision must next be made as to how to modify the process. If the improvement in color was significant, one or several repeats may then suffice. If inadequate, higher temperatures, longer times, and more or less strongly reducing conditions are next tried. If the reaction proceeds in a direction opposite to that desired, this enables deduction of what change of conditions is required. Some specimens behave in a contrary manner, turning blue on being heated in oxidizing conditions. It is suspected that certain specific iron-to-titanium ratios are involved in this behavior.

The surface of corundum is damaged at these very high temperatures, as has been reported previously even for the much lower temperature of jewelry repairing[23, 24]; the presence of chemicals can produce additional etching of the surface. This damage is one of the frequently seen signs of the use of heat treatment, as discussed in the identification section below.

If the stones are too dark, as with the 'inky' Australian sapphires, a lightening of the color may be possible. As described by Tombs[25, 26], these sapphires have not experienced the metamorphism of Sri Lanka sapphires and so require a different approach. In addition to some interesting speculations[21, 25, 26], he reports that temperatures as low as 1200 °C for just a few minutes may suffice. P. J. Kelly (unpublished observation) reports that 1000 °C used in the presence of a molten salt improves these dark sapphires. Some of these processes introduce a significant green component into the stones, others apparently do not.

An example of the inadvertent development of some blue color in a yellow stone during jewelry repair has been reported[24]. Heating sapphire to remove 'specks', implying either colored spots and/or silk, has long been practiced as, for example, described by Feuchtwangler in 1838[27]. For the sake of completeness it is worth mentioning Weigel's report of 1923[28], which appears to be the first detailed study on heat treatment of ruby, sapphire and spinel; he found that the blue of an Anakie sapphire gradually changed to yellow on heating to 1000–1300 °C and was a bluish green (o-ray, pale blue; e-ray, pale yellow–green) on returning to room temperature. This change was also included in Wild's listing of 1932, reproduced in *Table 3.1*.

Apart from the expected blue and yellow colors, some near-colorless corundums can be turned into an attractive, bicolor blue–yellow combination, while others give a range of reddish brown to pale, grayish brown (almost smoky colors, G. V. Rogers, unpublished observations). Some of these colors are illustrated in *Plate XV*, produced from near-colorless Sri Lanka corundum.

The recent absence of the once common brownish and purplish Thai rubies noted by R. Crowningshield (unpublished observation) indicates that such stones are now heat treated to remove the green or blue component which causes the off-color, as shown in *Plate XIV*, presumably by an oxidizing heating below 1000 °C; Wild in 1932 reported 450 °C for this change, as shown in *Table 3.1*, and Church also mentioned such a process in 1905[29]. Some such stones, however, require much higher temperatures and some do not seem to respond at all to treatment. If silk is present, then this too can be removed by process **C** at a high temperature, thereby intensifying the red by the absence of the white, light-scattering silk. A similar heat treatment has even been reported by R. E. Kane (unpublished information) to be used to improve flux-grown synthetic rubies. The products from all these treatments are stable. **E,F**

Other Treatments

When synthetic Verneuil-grown corundum is heated at 1600 °C or above, an additional effect is observed. This material contains curved striations derived from impurity concentration-fluctuations produced during the growth process[1]. Extended high-temperature heating permits diffusion to even out impurities, thus making the banding much less prominent and making a positive Plato test more difficult to observe[30]. **G**

Another high-temperature process uses irregular heating to produce small surface cracks which are then partially healed on additional heating, according to J. I. Coivula (unpublished experiments), particularly in the presence of a solvent such as borax. This can produce an appearance closely imitating the fingerprint inclusions seen in natural corundum; it has been seen in the Verneuil synthetic product[30]. **H**

Starting in 1979, a number of corundum gemstones have been seen which have their color concentrated in a thin surface layer. These have been various shades of orange[31, 32] and blue[32–34] faceted gems, as well as blue cabochons[35]. It was rapidly realized that these gemstones were produced by a process that had been patented many years ago for the enhancement of synthetic corundums and which is described in detail in Chapter 3, pp. 40–41. Extremely high temperatures are required for periods of many days to produce a depth of just a few thousandths of one inch (tens of micrometers). Since repolishing is required, the in-diffused layer is usually partially removed, thereby providing one of the ways of detecting this process, as further described below. In a variant (B. Cass, unpublished observation), the diffusion is carried out on the surface of a pale translucent cabochon and repolishing is employed to remove the surface layer on the upper portion of the cab. Such a cab has a good blue color when viewed from above, but looks quite pale when viewed from the side. The in-diffusion of cobalt into corundum to produce a sapphire-blue color has recently been reported to R. E. Kane (unpublished information); at the time of writing no samples have been analyzed to confirm this process. **J**

The overgrowth of a layer of synthetic corundum over a natural corundum, analogous to the Lechleitner process described above under beryl, has occasionally been performed on an experimental basis for ruby, both by the flux and hydrothermal processes; natural ruby has, at times, been employed as a seed for Verneuil growth[1], and Verneuil material has been used as a seed for flux growth[36]. These processes do not appear to have been used commercially. Claimed 'reconstructions', such as the early Geneva rubies[1], involved purified boule powder; more recent experiments that use ground-up natural sapphire as the feed material in a Verneuil torch still do not qualify as true reconstructions, as discussed at the end of Chapter 3. **M**

Dyed rubies and sapphires in which the color lies not in oil, but is deposited in cracks by the evaporation of a solvent or by chemical reaction, are possible; it may have been used, but has not usually been distinguished from the colored impregnations discussed below. **N**

The impregnation of rubies and sapphires with a colorless oil (or wax or plastic) to make fractures less prominent or to hide them completely has a history as long as the similar process discussed above under beryl. The technique there described applies equally well here, as does the discussion of Chapter 5, pp. 63–65. **O**

Oiling with a colored oil is occasionally seen and can be very deceiving. 'Ruby oil', in one instance consisting of mineral oil, a red dye, and a little disinfectant-type perfume, has been discussed in Chapter 5, pp. 65–66 and shown in *Figure 5.1*; *see also* treatment **K** under beryl, above. This type of treatment can, for example, be applied to naturally fractured, pale pink ruby to improve the color, or even to material that is almost colorless[37]. An old report[38] describes how a lavish diamond and ruby brooch was found to contain five synthetic rubies which had been heat crackled and then impregnated with a brown plastic. Another such report involved synthetic rubies that had holes drilled in them and then filled with plastic[39], clearly an early variant of a more recently patented process[40]. Both were said to give a very natural appearance! **P**

A variety of composite corundum stones have been made over the years, based on the various types described in Chapter 5, pp. 75–77. Typical is a ruby over a synthetic ruby doublet[41] or a layer of greenish natural sapphire over a synthetic ruby or blue sapphire back. An odd triplet was made of a top of synthetic sapphire attached by a dye layer to a faceted spinel back[42]. Also described have been star doublets, made of a base of star corundum covered by a crown of colorless or colored transparent material[43, 44]. A synthetic corundum triplet containing a diffraction grating as the center layer to imitate a star has ben patented[45]. The occurrence of ruby triplets containing a colored gel (*see Figure 5.7*) has recently been reported by R. E. Kane (unpublished information). **Q**

Colorless and colored reflective foils and other backings are today only rarely seen in corundums; the same is true of colorless corundum covered with a colored plastic coating. The plastic impregnations above could also be included here. The application of a red dye to the back of a stone was described by Cellini in 1568, as given in Chapter 2, p. 15. **R**

Asterism has been simulated in a patented process[46] in which a pattern of three sets of scratches at 120 degrees (i.e. 60 degrees) to each other are applied to (i.e. engraved onto) the back of a natural or synthetic corundum. In addition there are mirror backs, consisting of shiny metal foils with similar marks inscribed or embossed, which are attached to the back of a cabochon for the same purpose.

Identification

Identification begins with conventional gemological testing to establish the presence of corundum and to decide between the natural and the synthetic Verneuil, flux-grown, and Czochralski products[1]. Extra care must be taken now that the Verneuil banding may be made less prominent by process **G** and that **#**

fingerprint inclusions can be inserted, but only at or just below the surface, by process **H.** #

Synthetic overgrowth (so far experimental only) would be identified by the change in character on focusing down into the stone; a similar change in character is usually found in coated, doublet, or triplet composite gemstones. This type of examination also reveals asterism imitated in composite stones. Engraved stones, mirror-backs, and other crude star imitations should present no problems to the experienced eye. Further confirmation of all these possibilities follows conventional gemological practice.

Colored oiling can usually be recognized by the localization of the color in cracks seen on immersion in water or other fluids, or possibly on viewing by the light transmitted through a white translucent, plastic diffuser as with emerald in *Plate XXVII*. The spectrum of the dye rather than that of the corundum being examined is also an identifying characteristic.

Colorless oiling may present identification difficulties, particularly if the oil does not fluoresce. In the absence of a perfect refractive-index match, a very careful microscopic examination often reveals a faint outline of the filled cracks. An oily odor may be present, particularly when opening a parcel, and seepage may be seen on holding a hot point near the suspected region of oiled cracks. A solvent usually removes some of the oil and thus reveals the cracks in a 'slightly destructive' test.

Process **A**, the use of heat (or light) to remove the yellow fading color center (YFCC), involves such a low temperature that normally no evidence of its use remains. All the other heat-treatment processes **B** to **H** employ a sufficiently high temperature so that evidence of their use usually remains[30, 47]. One very positive piece of evidence is the presence of multiple girdles or unpolished pock-marked or 'burned' girdles or facets[30, 48], derived from the necessity for repolishing to remove the surface marks left by the very high temperatures and by borax or other substances used during heating; this repolishing is sometimes incomplete and gives the appearance shown in *Figures 7.6* and *7.7*. Stress fractures, that is cracking

Figure 7.6 Pock-marked facet on a heat-treated corundum; this facet had been missed during repolishing. Photograph by courtesy of the Gemological Institute of America

around inclusions due to differential thermal expansion, are frequently present; # this is described in Chapter 3, pp. 29–30 and *Figure 3.3*. Disk-shaped stress fractures are often seen surrounding tiny inclusions, as shown in *Figure 7.8*. Only in exceptionally clean stones is this characteristic absent.

Where yellow-to-brown-to-orange–pink sapphire is concerned, the first possibility to be eliminated is the presence of the yellow fading color center (YFCC) produced by process **K**. As suggested long ago by Crowningshield[49], a fading test by exposure to light may be needed; no routine gemological test is known at the time of writing which can distinguish the YFCC from the YSCC and other stable

Figure 7.7 Pock-marked multiple girdle on a heat-treated corundum; part of the girdle had been missed during repolishing. Photograph by courtesy of the Gemological Institute of America

yellow corundum colors, both in the yellow as well as in the padparadscha material. The deep yellow or golden-color sapphire produced by process **A** is stable to light; it frequently can be identified merely by its unusual color and the evidence of the use of high temperatures; additional clues, although in themselves not definitive, may be the absence of the spectroscopic iron line at 450 nm and subdued fluorescence[9]. J. I. Koivula (unpublished observations) has noted that rutile needles frequently melt during very high temperature heating and the diffusion of impurities from these molten drops can lead to diffuse orange–yellow halos, as seen in *Plate XVI*.

Several reports have dealt with the identification of sapphires where the blue color has been enhanced[30, 18, 47]. In addition to the evidence of the use of high temperatures (stress fractures and abnormal girdles and facets), there is the absence of the 450 nm iron line seen in the spectroscope, an abnormal dichroism (violet–blue and greenish- or grayish-blue), the frequent presence of a dull, chalky green fluorescence, the absence of silk, and the presence of abnormal cross-hatched and diffuse or blotchy color banding[30, 34]. Since treatment processes are occasionally modified, it is not always possible to rely on any one of these clues individually; as in much gemological testing, it is the preponderance of the evidence that counts.

The removal of small amounts of silk by process **C** from ruby or sapphires and the removal of a brown or purple component from ruby by process **F**, illustrated in *Plate XIV*, generally do not require quite as high temperatures (or rather can be performed in a reasonable length of time at lower temperatures) and so may not

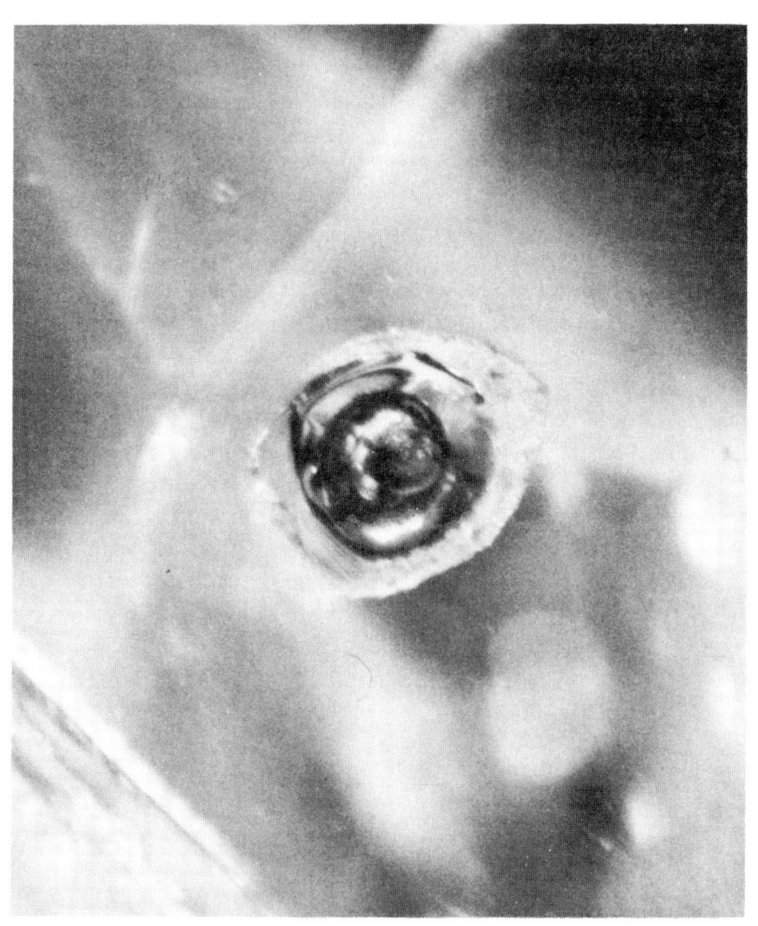

Figure 7.8 Disk-shaped crack surrounding an inclusion in a heat-treated corundum. Photograph by courtesy of the Gemological Institute of America

leave any evidence. The same is true of the development of potential asterism by process **D.**

The possibility that the banding of Verneuil growth may have been reduced and/or fingerprints induced by processes **G** or **H** must be kept in mind, once evidence of high-temperature heat treatment has been observed.

The in-diffusion of colors (only blue and orange have been seen at the time of writing, but all corundum colors are possible) and of asterism requires extended time at the very highest of temperatures and usually shows evidence of heating. Several additional phenomena help in the identification of in-diffusion[30, 48]. Immersion in water or other fluids in a cell held a few inches above a white sheet of paper usually reveals the concentration of color at the surface and, more particularly, its uneven distribution, derived from the necessary repolishing. This may completely remove the color layer from some of the facets or from around the girdle in a cabochon, as can be seen in *Plate XX*. If the in-diffusion has been performed on a preform, then the color is more likely to be present at the facet junctions. Also diagnostic is the 'bleeding' of color around open pits, fissures, and fractures, where impurities may have induced melting by locally lowering the melting point, and where the polishing was not able to reach the resulting deeply colored surface layer; this type of appearance is seen in *Plate XIX*. Stones with surface-diffused asterism show these same characteristics; they also show unnaturally strong stars.

One final test which may be mentioned derives from an apparent small change in the internal stresses that frequently remain after a very high-temperature heat treatment. It appears that this leaves the stones in a slightly more brittle state, so that they show more paper wear; another consequence also reported to R. Crowningshield (unpublished information) is this: when loose stones are permitted to dribble through the fingers onto a hard surface, the unheated stones go 'plink' while heated stones go 'plonk'.

It should be noted that the finding in the examination of a corundum that a high-temperature heat treatment has been used does not, in itself, prove that the material was originally 'worthless' and that the color was developed by the heating process. It is, in fact, quite possible that the color of the stone was originally quite close to its final color, and that only a small improvement was being sought or even that an enhancement in the color was not feasible!

References

1. K. Nassau, *Gems Made by Man,* Chilton, Radnor, PA (1980)
2. R. Crowningshield, Padparadscha: What's in a name?, *Gems Gemol.,* **19,** 30 (1983)
3. M. G. Townsend, Visible charge transfer band in blue sapphire, *Solid State Commun.,* **6,** 81 (1968)
4. G. Lehmann and H. Harder, Optical spectra of di- and trivalent iron in corundum, *Amer. Min.,* **55,** 98 (1970)
5. K. Schmetzer and H. Bank, Explanations of the absorption spectra of natural and synthetic Fe- and Ti-containing corundums, *Neus. Jb. Miner. Abh.,* **139,** 216 (1980)
6. K. Schmetzer and H. Bank, Die Farbursachen und Farben der Mineralart Korund, *Z. dt. Gemmol. Ges.,* **30,** 152 (1981)
7. K. Nassau, Heat treating ruby and sapphire: technical aspects, *Gems Gemol.,* **17,** 121 (1981)
8. K. Schmetzer, G. Bosshart, and H. A. Hanni, Naturally-colored and treated yellow and orange–brown sapphires, *J. Gemm.,* **18,** 607 (1983)
9. P. C. Keller, The Chanthaburi-Trat gem field, Thailand, *Gems Gemol.,* **18,** 186 (1982)
10. C. F. Fryer (Ed.), Sapphire, *Gems Gemol.,* **18,** 47 (1982)
11. C. F. Fryer (Ed.), Sapphire, color restoration, *Gems Gemol.,* **19,** 117 (1983)
12. R. Crowningshield, X-ray bombarded sapphires, *Gems Gemol.,* **13,** 57 (1969)
13. J. H. Oughton, Another color-changing sapphire, *Austral. Gemm.,* **11,** 4 (1972)
14. R. Crowningshield, Corundum observations and problems, *Gems Gemol.,* **16,** 315 (1980)
15. C. A. Schiffmann, Unstable colour in a yellow sapphire from Sri Lanka, *J. Gemm.,* **17,** 615 (1981)

16. F. H. Pough and T. H. Rogers, Experiments in X-ray irradiation of gem stones, *Amer. Min.*, **32,** 31 (1947)
17. K. Nassau, *in* Editorial forum, *Gems Gemol.*, **18,** 109 (1982)
18. H. S. Gunaratne, 'Geuda Sapphires' – their colouring elements and their reaction to heat, *J. Gemm.*, **17,** 292 (1981)
19. E. A. Jobbins, Heat-treatment of pale blue sapphire from Malawi, *J. Gemm.*, **12,** 342 (1971)
20. G. A. Tombs, Heat treatment of Australian blue sapphires, *Austral. Gemm.*, **13,** 186 (1978)
21. G. A. Tombs, *in* Editorial forum, *Gems Gemol.*, **18,** 43 (1982)
22. J. S. D. Abraham, Heat treating corundum: The Bangkok operation, *Gems Gemol.*, **18,** 79 (1982)
23. R. T. Liddicoat, Jr, The effect of the use of borax in repairs of corundum jewelry, *Gems Gemol.*, **14,** 342 (1974)
24. C. F. Fryer (Ed.), Dangers of heating sapphires during jewelry repair, *Gems Gemol.*, **18,** 106 (1982)
25. G. A. Tombs, Further thoughts and questions on Australian sapphires, their composition and treatment, *Austral. Gemm.*, **14,** 64 (1980)
26. G. A. Tombs, Heat treatment of Australian blue sapphires, *Z. dt. Gemmol. Ges.*, **31,** 41 (1982)
27. L. Feuchtwangler, *A Treatise on Gems,* p. 72, New York (1838)
28. O. Weigel, Über die Farbenänderung von Korund und Spinel mit der Temperatur, *Neues Jb. Miner., Beil.*, **48,** 274 (1923)
29. A. H. Church, *Precious Stones,* p. 66, 2nd Edn, London (1905)
30. R. Crowningshield and K. Nassau, The heat and diffusion treatment of natural and synthetic sapphires, *J. Gemm.*, **17,** 528 (1981)
31. R. Crowningshield, Some sapphire problems, *Gems Gemol.*, **16,** 194 (1979)
32. R. Crowningshield, Corundum observations and problems, *Gems Gemol.*, **16,** 315 (1980)
33. C. F. Fryer (Ed.), Induced surface coloration of natural sapphires, *Gems Gemol.*, **17,** 46 (1981)
34. H. A. Hänni, Zur Erkennung diffusionbehandelter Korunde, *Z. dt. Gemmol. Ges.*, **31,** 49 (1982)
35. C. F. Fryer (Ed.), Sapphire, diffusion colored, *Gems Gemol.*, **18,** 173 (1982)
36. R. T. Liddicoat, Jr, Flux synthetic ruby on a flame-fusion synthetic seed, *Gems Gemol.*, **15,** 174 (1976)
37. R. Crowningshield, Treated corundum, *Gems Gemol.*, **13,** 285 (1971)
38. Anon, Heat crackled synthetics, *Gemmologist,* **22,** 90 (1953)
39. R. T. Liddicoat, Jr, Doctored synthetic rubies, *Gems Gemol.*, **13,** 275 (1971)
40. G. B. Kitchel, US Patent 3 835 665, Apr 13 (1973)
41. C. F. Fryer (Ed.), Ruby and synthetic ruby assembled stone, *Gems Gemol.*, **19,** 48 (1983)
42. R. T. Liddicoat, Jr, Odd triplet, *Gems Gemol.*, **12,** 252 (1967–1968)
43. R. Crowningshield, Unusual doublet, *Gems Gemol.*, **12,** 305 (1968)
44. R. T. Liddicoat, Jr, Star doublets, *Gems Gemol.*, **12,** 280 (1971)
45. J. G. Donadio, US Patent 3 088 194, May 7 (1963)
46. R. S. Mukai, US Patent 2 511 510, Jun 13 (1950)
47. C. R. Beesley, Detection of heated sapphires, *Jewelers' Circ.-Keyst.*, **1982,** 106 (Aug, 1982)
48. C. F. Fryer (Ed.), Sapphire, heat treated, *Gems Gemol.*, **18,** 231 (1982)
49. R. Crowningshield, Developments and highlights, *Gems Gemol.*, **9,** 294 (1959)

Crystal, Rock, *see* quartz

Cyanite, *see* kyanite

Danburite

A pink color has been noted to form on irradiation (C. L. Key, unpublished observation). One would expect this to bleach on being heated; fading in light is also a possibility.

Deer Horn, *see* ivory

Diamond

Summary of Treatments

Irradiation:

A Produce green, blue, brown, yellow, black, and intermediate colors; all [s or u*], some [r]

B Pink changed to brown [f, r]

Heat:

C Modify irradiation colors to produce green, brown, orange, yellow, pink, red, purple, etc.; all [s], some [r]

D Modify natural yellow color [s, r?]

E Change color in chameleon diamond [f, r]

F Alter the surface by burning [s]

Other:

G Synthetic overgrowth [u†]

H Lasering [s]

J Coating and surface staining [u, f]

K Foils, composite stones, etc. [s or u]

* Unstable only in the sense that in some stones the color is only on the surface and may be lost on recutting or repolishing.

† If practical, it would be unstable only in the sense that it may be lost on recutting or repolishing.

Diamond is the cubic form of elemental carbon. It is one of the most thoroughly studied of substances, yet where its optical properties are concerned, particularly the causes of the colors, we must quote the words of Gordon Davies[1]: '. . . as yet, there is very little understanding of the optical properties from first principles'. Davies in his lengthy article gives a masterly summary reviewing the known data on diamond and clarifying many earlier misconceptions; this is essential reading, albeit at a high technological level. Using only a small fraction of the available literature, he nevertheless cites 297 references and his table of absorptions includes 38 lines and bands in the visible region between 400 and 700 nm! Other comprehensive technical works on diamond include the volumes by Berman[2] and Field[3].

There are many books written on the gemological and related aspects of diamond; one of the most useful ones is that by Bruton[4]. It must be emphasized that on the subject of nitrogen, 'platelets', and color in general, all these texts are essentially out of date and only Davies[1] should be followed, or the brief summary that is presented here.

Diamonds are usually divided into four types, designated Ia, Ib, IIa, and IIb; intermediate types also occur. Type Ia is sometimes subdivided into IaA and IaB. (Bruton[4] also lists type III for lonsdaleite, but since this is a hexagonal form of carbon it is definitely not diamond, a term strictly limited to the cubic form of carbon.)

Type I diamonds contain nitrogen as an impurity; this is typically present at the 0.1 percent level. Over 98 percent of clear, sizable natural diamonds are type Ia, in which the nitrogen is present as pairs of atoms (IaA) or as larger clusters containing an even number of nitrogen atoms (IaB). Type Ia diamonds also contain platelets, small flat inclusions about 100 nm across, which do *not* contain a significant amount of nitrogen[1]. Type Ib diamonds are very rare (less than 0.1 percent) and contain isolated nitrogen atoms.

All type I diamonds absorb in the infrared region from 6 to 13 μm and in the ultraviolet beyond 300 nm; they usually show a blue fluorescence. They are frequently yellowish, showing the 'cape series' of absorption lines in the spectroscope, as in *Figure 7.9(a)* and *(b)*; this series of lines is derived from 'vibronic' interactions. The cape series includes the 415 nm line, designated 'N3' and

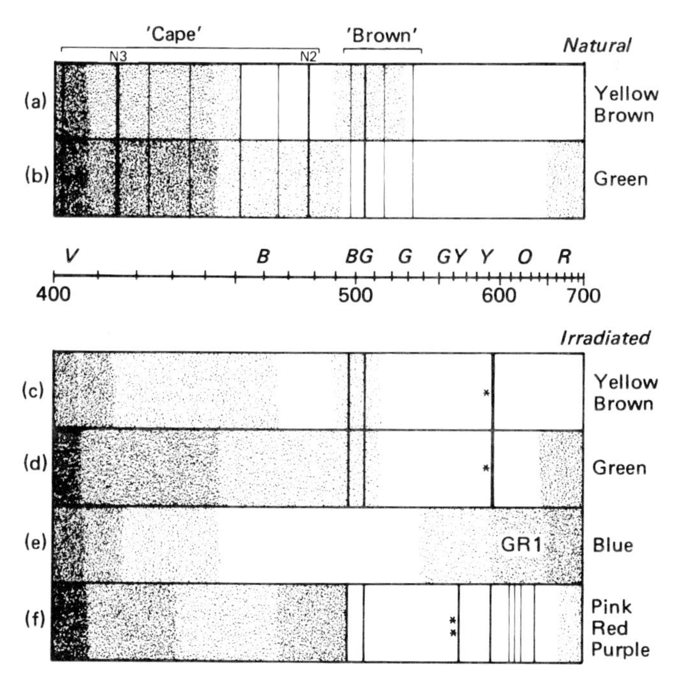

Figure 7.9 Representative absorption spectra of diamond as seen in a hand spectroscope, shown in idealized form: * indicates variable intensity, ** indicates a line seen both in fluorescence and in absorption. Note that the 'cape' and 'brown' series of lines can occur independently of each other and are frequently also present in irradiated stones, derived from the original color; they are omitted there for clarity

probably derived from three nitrogen atoms surrounding one carbon atom in a flat configuration, the weaker 'N2' line at 478 nm, and other yet weaker lines. Diamonds with brownish rather than yellowish tints show a weak line at 504 nm accompanied by other weaker lines[5, 6] also shown in *Figure 7.9(a)* and *(b)*; some of these brownish stones have a greenish fluorescence. Nitrogen has one electron more than carbon and is therefore a donor of electrons. It does not produce electrical conductivity because the donor energy is too large, resulting in a deep donor and insulating properties.

Type II diamonds do not contain nitrogen. They have no characteristic absorptions in the 6 to 13 μm region and transmit in the ultraviolet below about 300 nm; they are excellent conductors of heat. Type IIa is restricted to the colorless stones that contain no nitrogen or other significant impurities and are transparent in the ultraviolet to about 225 nm. As pointed out by Davies[1,7], this group continues to shrink as nitrogen detection equipment becomes more sensitive; he recommends that it be eliminated. Type IIb diamonds are blue because of their boron content (earlier reports of aluminum cannot be supported[1,7]; they are rare, with an occurrence of about 1 percent. These IIb diamonds transmit in the ultraviolet to about 250 nm and usually show phosphorescence under short-wave ultraviolet radiation. Although most type IIb diamonds are blue, there are some brown and gray ones as well. Boron, having one electron less than carbon, is an acceptor; the acceptor energy is very low, leading to a shallow level and electrical conductivity.

Synthetic diamond exactly parallels the natural, with nitrogen addition producing an insulating yellow and with boron giving a blue that conducts electricity[7].

A Irradiation is used frequently to alter the color of diamonds, usually to produce from off-color stones a wide range of attractive blue, green, brown, orange, black (very dark green), and yellow colors, as seen in *Plate X*. There are two major processes, one using electrons, the other using radium salts, neutrons from a nuclear reactor, or protons, deuterons, alpha particles, or other heavy particles from cyclotrons or linear accelerators (*see* Appendix B). These processes are performed commercially (*see* Appendix D), but relatively little information is released so that the following description is of necessity incomplete. It is usually assumed, and a very reasonable assumption it is, that the irradiation-produced colors derive from color centers. To date no actual details have, however, been clarified and nothing more than this vague statement is justified!

Irradiation by burial in a radium salt for up to one year was used about 1904 by Sir W. Crookes. He found that a green color was produced, but that radioactivity was also induced in the diamond[8]; this process is now known to involve the implantation into the surface of the diamond of fast-moving nuclei recoiling from their disintegration. Radioactive stones produced by this radium process are still seen occasionally, sometimes with extremely high levels of radioactivity[9]. Although very high temperatures may remove the green color, the radioactivity still remains (despite some erroneous statements in the literature to the contrary); *see also* Chapter 4, pp. 58–60. Lind[10] in 1923 showed that it is the alpha particles that produce the coloration.

Subsequently, many types of irradiation have been tried, including alpha particles, protons, deuterons, and so on from cyclotrons or linear accelerators and neutrons from a nuclear reactor. All give essentially the same result. A wide absorption band, usually called the 'GR 1' band is formed, extending from the 'GR 1' line in the infrared at 741 nm well into the yellow–green region, as shown in *Figure 7.9(e)*. This can produce a green, blue–green, dark green, or a black color. Even with the use of cooling water, the heat generated within the stone during the irradiation often produces additional color changes, as discussed below under process **C**.

A

Except for the rarely used but deeply penetrating gamma rays, which are not as effective in producing the coloration, and the at present most frequently used neutrons, these radiations have a relatively short range in diamond (*see also Tables B.1* and *B.2* in Appendix B). As a result, the coloration is localized and this can be recognized. When irradiated in a cyclotron from the pavilion side, a peculiar umbrella-shaped concentration of color surrounding the culet is seen when looking down through the table, as shown in *Figure 7.10.* If treated from the crown side, the localization of color is visible in various positions[11], including the appearance of a dark ring when viewing the stone from the pavilion side while it lies on a white surface. Irradiation has been performed on rare occasions from the side[11, 12].

Figure 7.10 'Umbrella-shaped' marking surrounding the culet on a diamond irradiated from the pavilion direction. Photograph by courtesy of the Gemological Institute of America

The second major irradiation technique employs high-energy electrons. Here too a large amount of heat is created and cold running water is used to prevent the simultaneous occurrence of a heat treatment. Electrons have a much deeper penetration than heavy particles, as shown in *Table B.2* of Appendix B, and so produce a much more uniform coloration. The product has a blue or blue–green color, depending on the material and on the treatment circumstances; lower energy electrons, such as a fraction of one eV, tend to produce a blue color but do not penetrate as deeply as electrons of several eV which, however, tend to produce a blue–green. Here too the dominant initial product is the 'GR 1' band. Details and references on early work are given by Michel[13] and some data on the irradiation intensities and times used are given by Gübelin[14].

Neutron and electron irradiation are today the preferred irradiation techniques. Both produce uniform coloration because of their good penetration. All the colors so produced are stable.

At times unexpected colors are formed. Thus, one diamond that was examined gemologically before and after irradiation showed a greenish yellow fluorescence and was therefore expected to treat to a good 'chartreuse' color. Instead, neutron irradiation produced an orange–red stone with a bright orange–red fluorescence; absorption spectra were given[15]. One suspects that this may have been a stone having significant Ib characteristics, as discussed below. Another unexpected change was in a cloudy diamond (possibly a IIb?) irradiated with neutrons which turned an unusual, rich sky-blue color[16]. The question is, of course, often asked whether one could turn a semiconducting type-IIb stone blue or a deeper blue so that it would seem to be a natural color because of its conductivity. However, at least one such stone lost its conductivity on being irradiated[17]. The heat sometimes applied to a diamond during jewelry repair could affect the irradiation-induced colors, as described in process **C** below; the temperatures involved in steam cleaning are sufficiently low to present no such problems. **A**

Several times it has been observed that some natural pink diamonds turn colorless, or a brown or purple color on being exposed to very small doses of X-rays[18–20], at least irradiation doses small compared to those employed in process **A**. These color changes are due to very fleeting color centers which return to the original color on exposure to light and/or on gentle warming, such as while being illuminated by a strong lamp. **B**

Heating is widely used to modify the colors produced by the irradiations described under process **A** above. In general, the change is one from blue to green to brown to yellow and back to the original color. Depending on the nature of the stone and the conditions used, this sequence may stop at various points and then may not proceed further. The color cannot be made any lighter than that of the stone originally, but may remain darker. A second irradiation may produce reversion to the blue or green, but the result may vary depending on the nature of the intervening heat treatment. If the original diamond was of the rare type Ib, then irradiation and heating is reported to produce pink, red, mauve, or purple colors[1,4]. However, it has been noted (R. Crowningshield, unpublished observations) that irradiated stones of these colors show a strong orange fluorescence, a property also present in the extremely rare natural pink stones, thus raising the possibility that the color of very pale pink natural stones is being intensified by the irradiation. The absorption is shown in *Figure 7.9(f)*. **C**

Collins[21] has shown in detail how heating destroys the 'GR 1' absorption, starting at about 400 °C, with the formation or intensification of the 497, 503, and 595 nm* absorption lines, which become strongest at about 800 °C, producing from the blue a green, yellow, orange, or brown color depending on the intensity. Typical spectra are shown in *Figure 7.9(c)* and *(d)*. By 1000 °C, however, the 595 nm line completely disappears without affecting the color significantly; accordingly, it is not by itself a suitable test for irradiation in a yellow-to-brown diamond. One wonders if the bright yellow irradiated stone with a very weak 592 nm absorption recently seen[22] may not have been treated in this way. In natural yellow diamonds the 504 nm absorption line usually seems to be stronger than that at

* This line may be listed as 592 nm, 595 nm, etc. The exact position of such lines may depend on the temperature as well as on the observation technique.

497 nm, while after irradiation and heating these two lines may be equal or the 497 nm line may dominate[5]; however, these changes cannot always be relied upon (Crowningshield, unpublished observations). The 'cape' absorption lines are not affected by irradiation and heating and can give an indication of the original color of the diamond before treatment. The 'brown' absorption lines may be altered by heating. **C**

Heating under high pressure for just a minute or so to temperatures near 2000 °C has been shown to change type Ia diamonds to a bright yellow, with a change in the platelet structure as summarized by T. Evans (pp. 419–424) in Field[3]. Conversely, type Ib diamonds have been changed to type IaA with a lightening of the yellow color[23], and type Ia diamonds were changed to type Ib under slightly different conditions[24]. The possible commercial significance of these experiments regarding the decolorizing of natural or synthetic yellow diamonds is not yet clear. **D**

A special group of diamonds showing unusual color changes on heating are the 'chameleons'. Most commonly these are grayish green to yellowish green stones which turn a bright yellow 'fancy' color on being kept in the dark or on being warmed slightly with an alcohol flame or the like[17, 25]. On exposure to light for a few minutes or to ultraviolet light for just seconds, the original color returns. A very shallow color center (*see* Chapter 4, pp. 53–56) is clearly involved; this forms in the light, but even room temperature in the dark is enough to fade the color. Also reported[20] has been a salmon-pink diamond, which similarly turned a bright red color on heating (while being boiled in acid) and returned to salmon pink within a few days in light. Other color changes may be possible, but generally have not been investigated. **E**

Diamond, being carbon, burns on being heated to a sufficiently high temperature. This has happened occasionally during jewelry repairing[17, 26]. A superficial burning can leave a thin, whitish crust on the surface, and some rough diamonds have been so treated to make them seem much whiter while being purchased than later after having been polished[27]. **F**

The overgrowth of diamond on natural diamond to increase its weight has been frequently claimed. Several techniques have been described, including low-pressure vapor-phase growth, patented by Eversole in the USA in 1962 and by Derjaguin and Spitsyn in the Soviet Union in 1956. More recently, an ion-implantation process by Freeman and co-workers was granted two patents in 1977 in Great Britain. Not usually stated are the actual overall growth rates, which are at most 1/50 mm per week for the low-pressure process and perhaps as much as 1/2 mm per week, but over a small area only, in the very difficult and expensive ion-implantation process. Clearly, these processes are totally impractical at present. Since it is possible that a layer of blue diamond could be added, future developments are worth watching, although the prospects are dimming with so little improvement after so many years of intensive investigation. Details and references have been given elsewhere[7, 28]. **G**

Laser drilling has been used to open passageways to inclusions since about 1970[29]; the inclusions can then be vaporized by the laser or etched out with acid. The clarity grade may not be improved by such a treatment, but the saleability may[29]. The hole may range from 25 μm (one-thousandth of one inch) in diameter down to less than one-tenth of this size; there is no practical limit to the depth. The **H**

H

hole is usually drilled perpendicular to a nearby facet, so that the direction error originating from refraction is avoided. Natural cavities in diamond which superficially look like laser holes have been observed[17]. Another use of lasers is in the system recently introduced by the Gem Trade Laboratories of the Gemological Institute of America for marking certificate numbers and other information on the girdle of diamonds. Illustrations of both of these 'lasered' products are given in Chapter 5, *Figures 5.5* and *5.6*.

J

Coatings on diamonds are seen occasionally; they are used to simulate a fancy color or to make a diamond appear white, usually by adding a complementary pale blue or purple absorption to neutralize an off-color yellow absorption. Diamonds have been coated with pink nailpolish, as in the 1983 Sotheby Park Bernet substitution described in Chapter 5, p. 72, and in other pink[20] and blue[30, 31] coating reports. Thin sputtered coatings, usually composed of fluorides as seen in the purple anti-reflection 'bloom' on camera lenses, have also been employed, as well as silica coatings. Boiling with acids or the use of a mild abrasive, such as the quartz in an eraser, removes the coating. Thin, colored enamel coatings have been fused onto the surface of diamond; these can be quite hard and a sapphire point may be required to scratch them.

The application of colored stains to the surface of a diamond has long been practiced, again using purple or blue colors such as potassium permanganate solution, indelible pencil, or marking ink! Merely applying a tiny amount to the girdle under one of the prongs can 'improve' the apparent color of a yellowish diamond significantly. A detailed study on various ways of detecting coatings and surface stains has been published[32]. Care must be taken not to confuse coatings with stains derived from water having a high iron content[33, 34].

K

Foils to reflect light and modify the color of diamonds used to be common at one time; historical references have been given in Chapter 5, pp. 72–73. They have not been used much recently, except occasionally to provide a fake back to a flat, rose-cut diamond to make it appear a solid stone[35, 36] in what could be termed a diamond–air doublet. Doublets have been reported consisting of diamond-on-diamond and of a diamond crown cemented to a pavilion of quartz[16], white sapphire, or strontium titanate[36]; the last of these has been called a 'pavilion diamond' even though it is the crown that is the diamond! In these reports there has always been a continuing problem of separation because of the poor holding power of the cement.

Finally, mention can be made of one of the early apparent myths that diamond is softened by goat's blood. As discussed in Chapter 2, pp. 9–10, this was an actual observation and makes perfect sense once it is understood that the 'diamond' of that period was in fact quartz and that the 'softening' process involved cracking the stone to make it receptive to a colorant.

#

The identification of foil backs, doublets of the various types, laser drilling, coatings, and surface stains is based on a careful examination under magnification; various ways of detecting coatings have been described by Miles[32]. Synthetic overgrowth is not practical at the time of writing.

The distinction between natural colored diamonds and those produced by irradiation combined with heating '. . . is a matter for the specialist' (Anderson[5], p. 238); some details are given in the textbooks[5, 6] and in a recent review by Scarratt[37].

Enhancement by simple irradiation

Plate I Pearls in the natural state (below) and irradiated (above); the dark colour is stable to light

Plate II Kunzite in the natural pink state (left) and irradiated (right); the green color fades rapidly in light

Plate III Irradiated Maxixe-type beryl (center) and two originally matching stones, after 1 week in the sun (left) and after half-an-hour in boiling water (right)

Enhancement by simple irradiation

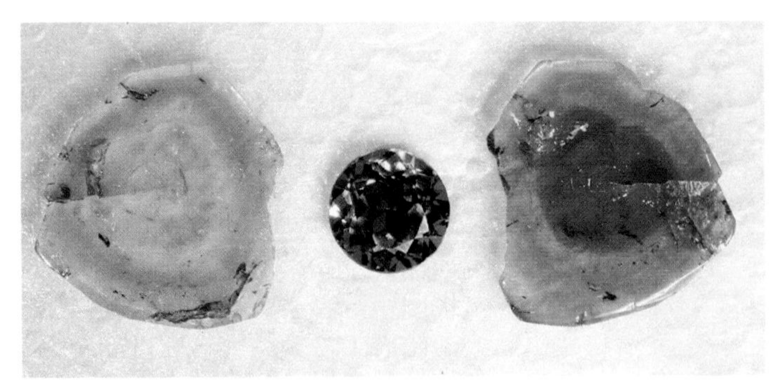

Plate IV Almost colorless tourmaline turned a peach color by irradiation (center) and originally matching watermellon slices, one irradiated (right); colors are stable to light

Plate V Untreated cats-eye tourmaline (center), with two originally matching stones, one of which turned yellowish on irradiation (left) and the other a deeper reddish (right); colors are stable to light

Plate VI Pale blue tourmaline (center) surrounded by originally matching specimens with colors changed by irradiation; colors are stable to light

Enhancement by irradiation plus heating
(those illustrated are stable to light)

Plate VII Section from an amethyst crystal which was treated by heat and irradiation to turn amethyst–citrine bicolor

Plate VIII Faceted amethyst–citrine, weighing 108.5 carats. Photograph by courtesy of the Gemological Institute of America; copyright with *Gems and Gemology*, used with permission

Plate IX Four colorless quartzes, all irradiated to turn them smoky (left and right) and two heated to produce greenish yellow quartz (below) and finally colorless again (above)

Enhancement by irradiation plus heating

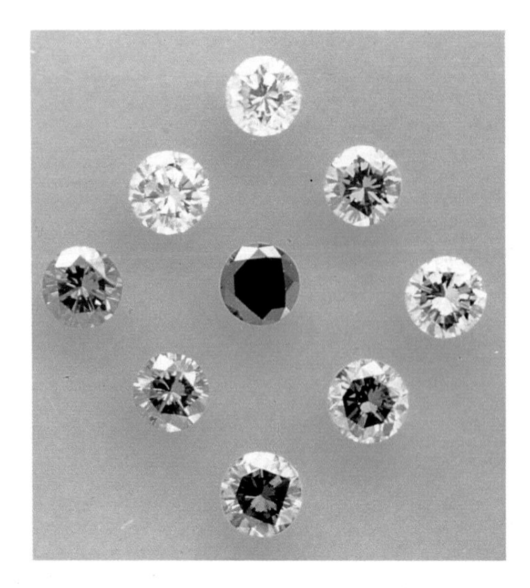

Plate X Originally near-colorless diamonds, some irradiated and some heated; colors are stable to light. Photograph by courtesy of the Gemological Institute of America

Plate XI Above: initially matched pale green aquamarines, both heated to become blue and one then irradiated to return it to green; colors are stable to light
Below: pink Maxixe-type beryl, a matching gemstone irradiated to turn it blue, and a greenish gemstone similarly turned dark blue–green; the blue and green colors fade in light

Plate XII Initially colorless topaz gemstones turned various colors by irradiation; the orange and brown ones fade in light, the blue ones are stable. The blue stones have been heated to remove the brown color; the blue colour is stable to light

Enhancement by simple heating
(those illustrated are stable to light)

Plate XIII Almost colorless 'geuda' sapphire heated to become a deep blue. Photography courtesy of the Gemological Institute of America; copyright with *Gems and Gemology*, used with permission

Plate XIV Brownish Thai ruby (left) with a similar gemstone which has been heat treated to remove the off-color (right). Photographs by courtesy of the Gemological Institute of America

Plate XV Some of the colors obtained by heat treatments of initially almost colorless sapphires by G. V. Rogers; note the yellow–blue bicolor at the lower right.

Enhancement by simple heating
(those illustrated are stable to light)

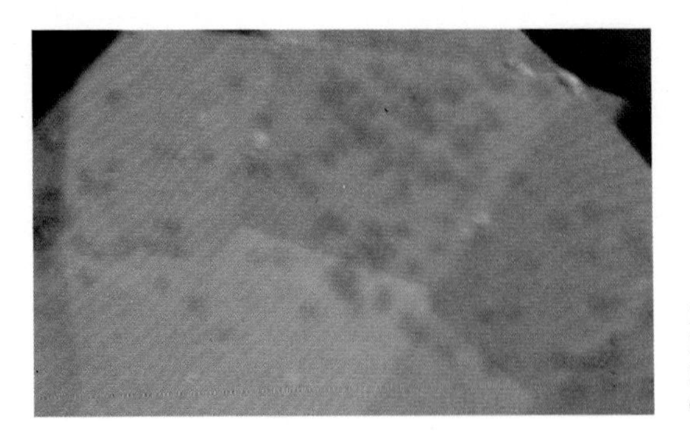

Plate XVI Orange–yellow halos as observed in some heat-treated corundums by J. I. Koivula; magnification 25X. Photograph by courtesy of the Gemological Institute of America

Plate XVII Initially matched dark amethysts heated by D. McCrillis to become paler (above) and by J. Call to become colorless and milky (below)

Plate XVIII Initially matched pale brown tanzanites, one heated by J. Call to become blue

Enhancement by heating with additives
(those illustrated are stable to light)

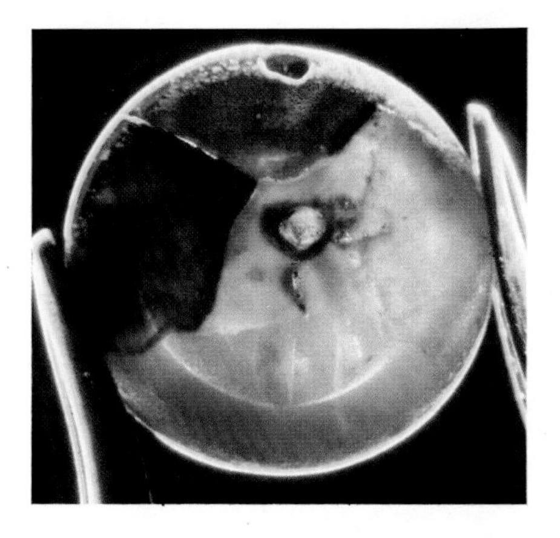

Plate XIX 'Bleeding' of color on the back of a color-diffused sapphire cabochon. Photograph by courtesy of the Gemological Institute of America

Plate XX Localization of color in an orange–red, color-diffused sapphire revealed by immersion in methylene iodide. Photograph by courtesy of the Gemmological Institute of America; copyright the *Gems and Gemology*, used with permission

Plate XXI Amber bead containing stress fractures produced by heating in oil. Photograph by courtesy of the Gemological Institute of America

Enhancement by other processes

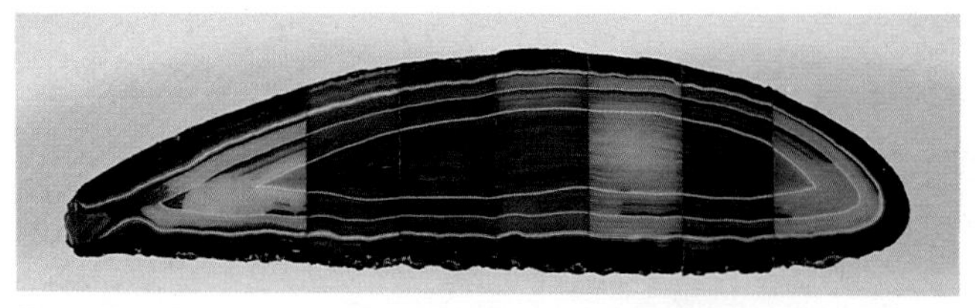

Plate XXII Slab of agate sliced and dyed various colors in Idar Oberstein. Photograph by courtesy of J. S. White, Smithsonian Institution

Plate XXIII Dyed 'crackled' quartz; material by courtesy of the Gemological Institute of America

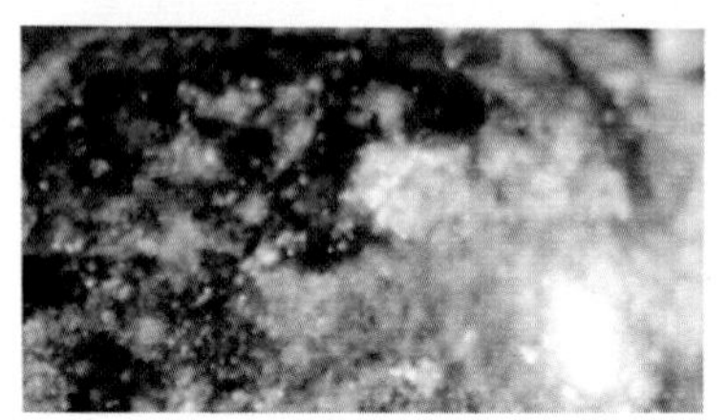

Plate XXIV Sugar-treated opal, magnification 25X; color is stable to light. Photograph by courtesy of the Gemological Institute of America

Plate XXV White, opaque opal (center below) and originally matching material impregnated with colorless plastic to show intense play-of-color; colors are stable to light

Plate XXVI Faceted colorless quartz made to imitate alexandrite and citrine by surface coatings applied to the back

Plate XXVII Whitish beryl dyed green with green and yellow dyes, in dark field illumination (left) and in diffuse transmitted light (right). Photograph by courtesy of the Gemological Institute of America; copyright with *Gems and Gemology*, used with permission

Briefly, cyclotron irradiation produces a concentration of color either at the culet, as in the 'umbrella' of *Figure 7.10*, at the table, or on the girdle[11, 12]; the last of these must be carefully distinguished from the naturally occurring radiation stains on girdle naturals[38] (parts of the original diamond surface left unpolished); these are significant because natural diamonds usually only have a green color if part of the green skin is retained somewhere on the faceted stone[6, 39]. Recutting or repolishing may result in the loss of some or all of the color. Green diamonds may be checked for mossy green surface particles derived from irradiation, and for radioactivity derived from the use of radium salts (days of exposure on a photographic film may be required), and blue diamonds for electrical conductivity and ultraviolet transparency, both indicative of the natural blue IIb diamonds.

Other irradiation techniques do not produce such easy signs. The spectroscope is the most important identification technique, often used with cooling, either with an aerosol refrigerant-gas, or, preferably, with dry ice or with liquid nitrogen, to make certain absorption lines more prominent[40], such as the important one at 592 nm; this line also occurs in untreated diamonds but only very rarely and then only in the green patches at the surface. The relative intensity of the 497 and 504 nm lines has been discussed under process **C** above. Some details are given in *Figure 7.9*; considerable variation and additional anomalous absorption lines can occur. The fluorescence can also provide clues.

References

1. G. Davies, The optical properties of diamond, in *Chemistry and Physics of Carbon,* P. L. Walker, Jr, and P. A. Thrawer (Eds), Vol 13, pp. 1–143, Dekker, New York (1977)
2. R. Berman (Ed.), *Physical Properties of Diamond,* Clarendon Press, Oxford (1965)
3. J. E. Field (Ed.), *The Properties of Diamond,* Academic Press, New York (1979)
4. E. Bruton, *Diamonds,* 2nd Edn, Chilton, Radnor, PA (1978)
5. B. W. Anderson, *Gem Testing,* 9th Edn, Butterworths, London (1980)
6. R. T. Liddicoat, Jr, *Handbook of Gem Identification,* 11th Edn, Gemological Institute of America, Santa Monica, CA (1981)
7. K. Nassau, *Gems Made by Man,* Chilton, Radnor, PA (1980)
8. W. Crookes, *Diamonds,* pp. 108–110, Harper and Brothers, London (1909)
9. R. Crowningshield, Radium-treated diamonds, *Gems Gemol.,* **12,** 304 (1968)
10. S. C. Lind, The coloring of the diamond by radium radiation, *J. Franklin Inst.,* **196,** 521 (1923)
11. C. F. Fryer (Ed.), Cyclotron-treated diamonds, *Gems Gemol.,* **17,** 40 (1981)
12. R. T. Liddicoat, Jr, An unusual cyclotron-treated diamond, *Gems Gemol.,* **15,** 72 (1975)
13. H. Michel, *Die künstlichen Edelsteine,* W. Diebner, Leipzig (1926)
14. E. Gübelin, Farbe und künstliche Farbveränderung von Diamanten, *Neue Zürcher Zeitung, Beilage, Technik* (Oct. 12, 1960)
15. R. Crowningshield, Treated red–brown diamond, *Gems Gemol.,* **12,** 44 (1966)
16. R. T. Liddicoat, Jr, Banded diamond and diamond doublets, *Gems Gemol.,* **14,** 106 (1972–1973)
17. R. Crowningshield, Diamonds, *Gems Gemol.,* **15,** 57 (1975)
18. R. Crowningshield, Developments and highlights, *Gems Gemol.,* **9,** 360 (1959–1960)
19. L. B. Benson, Developments and highlights, *Gems Gemol.,* **10,** 45 (1960)
20. C. F. Fryer (Ed.), Pink diamonds, *Gems Gemol.,* **19,** 43 (1983)
21. A. T. Collins, Investigating artificially coloured diamonds, *Nature,* **273,** 654 (1978)
22. C. F. Fryer, Unusual absorption spectrum, *Gems Gemol.,* **18,** 229 (1982)
23. R. M. Chrenko, R. E. Tuft, and H. M. Strong, Transformation of the state of nitrogen in diamond, *Nature,* **270,** 141 (1977)
24. M. R. Brozel, T. Evans, and R. F. Stephenson, Partial dissociation of nitrogen aggregates in diamond by high temperature–high pressure treatments, *Proc. Royal Soc.,* **A361,** 109 (1978)
25. C. F. Fryer (Ed.), 'Chameleon' diamond, *Gems Gemol.,* **18,** 228 (1982)
26. R. T. Liddicoat, Jr, Diamond pyromania, *Gems Gemol.,* **14,** 89 (1972)
27. R. Crowningshield, Burned surfaces on rough diamonds, *Gems Gemol.,* **13,** 345 (1971)

28. K. Nassau and J. Nassau, The history and present status of synthetic diamond, *Lap. J.*, **32,** 76, 490 (1978)
29. R. Crowningshield, Laser beams in gemology, *Gems Gemol.*, **13,** 224 (1970)
30. R. Crowningshield, A painted blue diamond, *Gems Gemol.*, **15,** 124 (1975–1976)
31. R. Crowningshield, Follow-up on a blue–gray diamond, *Gems Gemol.*, **15,** 183 (1976)
32. E. R. Miles, Diamond coating techniques and methods of detection, *Gems Gemol.*, **10,** 355 (1962–1963)
33. R. Crowningshield, Diamond discolored by water, *Gems Gemol.*, **12,** 22 (1966)
34. R. Crowningshield, A diamond color mystery, *Gems Gemol.*, **15,** 182 (1976)
35. R. Crowningshield, Highlights, *Gems Gemol.*, **9,** 269 (1959)
36. R. Crowningshield, Foiled again!! *and* More on doublets, *Gems Gemol.*, **13,** 374 (1971–1972)
37. K. V. G. Scarratt, The identification of artificial coloration in diamond, *Gems Gemol.*, **18,** 71 (1982)
38. C. F. Fryer (Ed.), Natural radiation stains, *Gems Gemol.*, **17,** 41 (1981)
39. Yu. L. Orlov, *The Mineralogy of Diamond,* Wiley, New York (1973)
40. S. C. Hofer and D. V. Manson, Cryogenics, an aid to gemstone testing, *Gems Gemol.*, **17,** 143 (1981)

Disthene, *see* kyanite

Dolomite, *see* marble

Doom (Doum) Palm Nut, *see* ivory

Dravite, *see* tourmaline

Eilat Stone, *see* chrysocolla

Elath Stone, *see* chrysocolla

Elbaite, *see* tourmaline

Emerald, *see* beryl

Eyes, *see individual gemstone variety involved*

Feldspar Group

Including: adularia, albite, amazonite, aventurine feldspar, labradorite, microcline, moonstone, oligoclase, orthoclase, perthite, plagioclase, sunstone, etc.

Summary of Treatments

Heat:
- **A** Modify optical effect (e.g. moonstone) [s]
- **B** Remove color (e.g. amazonite) [s, r]

Irradiation:
- **C** Restore color (e.g. amazonite) [s, r]
- **D** Produce smoky and other colors [s? or u, f]

Other:
- **E** Foil, backing, coating, etc. [s or u]

The feldspars comprise two closely related families of minerals and gemstones. The potassium feldspars include sanidine, orthoclase, and microcline, all approximately $KAlSi_3O_8$. The plagioclase feldspars consist of a continuous chemical substitution series ranging from albite, $NaAlSi_3O_8$, to anorthite, $CaAl_2Si_2O_8$, with general formula $(Na,Ca)Al(Al,Si)Si_2O_8$; the intermediate members are oligoclase, andesine, labradorite, and bytownite. There is some solid solubility above about 660 °C between the potassium feldspars and albite; at lower temperatures these solutions separate into the layered perthite feldspars. Excellent summaries are found in specialized treatises[1,2], as well as in mineralogy textbooks such as that by Berry, Mason, and Dietrich[3].

The most important feldspar gemstone varieties include blue-to-green amazonite, a microcline containing a color center; labradorite with its varient spectrolite, both being labradorite plagioclases having layered structures[4] and showing interference between rays of light reflected from successive layers, producing almost any color of the rainbow depending on the layer spacing; moonstone (or adularia), being semitransparent albite or orthoclase showing a bluish sheen derived from an internal layer structure and/or light-scattering centers; and sunstone (or aventurine feldspar), an oligoclase containing included minerals such as hematite or copper[5].

Heating of the layered structure feldspars for extended periods to over 1000 °C generally causes the perthitic layers to homogenize into a single solid-solution by a diffusion process, and the material then usually becomes homogeneous and transparent. At intermediate temperatures, say 900 °C for several hours in air, the moonstone effect becomes less intense, but with a slightly more bluish appearance[6]. Similar effects can be expected in other perthites; labradorite and spectrolite might be affected in some way, but the author is not aware of any reports on this. The effect of heat on the various optical effects occurring in the feldspar gemstones does not appear to be known in great detail and may merit further study. **A**

The blue-to-green color of amazonite is lost on heating to over 300 °C[5,7]. The color can be restored by almost any type of irradiation if the heating was not carried to so high a temperature that water is lost. The color apparently involves both lead and water impurities, with irradiation producing a color center and heating producing its destruction. The color of some amazonite can be deepened by irradiation (G. Rossman, unpublished observation). It is possible that microcline containing both the necessary impurities exists in nature in a colorless form that could produce amazonite on irradiation. **B** **C**

Many feldspars, including sanidine, produce smoky colors on irradiation[5,8], although orange and green have also been reported for oligoclase[9]. While not specified in the reports, these colors might be unstable and could fade in light. **D**

Foils were widely used at one time but are now rarely seen. Bauer[10] states that a black backing heightens the optical effect in moonstone. A surface coating of paraffin wax is commonly applied, particularly to beads, cabochons, or tumbled stones of amazonite[11], which has a strong tendency to produce many small surface cracks derived from its incipient cleavage. **E**

References

1. W. A. Deer, R. A. Howie, and J. Zussman, *Rock Forming Minerals,* Vol 4, Wiley, New York (1963)
2. T. F. W. Barth, *Feldspars,* Wiley, New York (1969)
3. L. G. Berry, B. Mason, and R. V. Dietrich, *Mineralogy,* 2nd Edn, Freeman, San Francisco (1983)
4. H. C. Bolton, L. A. Bursill, A. C. McLaren, and R. G. Turner, On the origin of the colour of labradorite, *Phys. Status Solidi,* **18,** 221 (1966)
5. A. M. Hofmeister and G. R. Rossman, Color in feldspars, *Rev. Min.,* **2,** 271 (1983)
6. E. Spencer, A contribution to the study of moonstone from Ceylon and other areas and of the stability-relations of the alkali-feldspars, *Min. Mag.,* **22,** 291 (1930)
7. K. Przibram and J. E. Caffyn, *Irradiation Colours and Luminescence,* p. 253, Pergamon Press, London (1956)
8. F. H. Pough, The coloration of gemstones by electron bombardment, *Z. dt. Ges. Edelstkeink.,* **20,** 71 (1957)
9. F. H. Pough and T. H. Rogers, Experiments in X-ray irradiation of gem stones, *Amer. Min.,* **32,** 31 (1947)
10. M. Bauer, *Precious Stones,* Vol 2, p. 430, Dover Publications, New York (1968)
11. R. Crowningshield, Treated amazonite, *Gems Gemol.,* **11,** 102 (1963–1964)

Flint, *see* chalcedony

Fluorite

Including: blue john and fluorspar.

Summary of Treatments

Heat:
A Lighten blue and other colors [s]
Irradiation:
B Induce or darken colors [s or u, f]
Other:
C Colorless impregnation [s]
D Colored impregnation [s or u, f]

The blue to deep blue to black color seen in some fluorites is derived from a color center, the *F*-center (named after the German word 'Farbe' for 'color'). This involves an electron-center precursor, and a missing fluorine atom in the fluorite, CaF_2. An electron liberated by irradiation or other processes[1] from a hole-center precursor can be trapped at the fluorine vacancy to become the electron color-center. Many other color centers are known in fluorite, including those producing yellow, green, violet, pink, and other colors; some of these are stable to light, others are not. Details have been given in many places[1–4].

The color of most of these color centers can be lightened by a gentle heating to 100–150 °C and is totally removed at 200–300 °C. This process has long been used to lighten some excessively dark 'blue john', a deep blue, banded massive fluorite found in Derbyshire, England, and used for carved ornamental objects. **A**

Irradiation can restore the color of overheated 'blue john' or produce blue and other colors in colorless materials; these colors may be stable to light or may fade, depending on the material. **B**

Plastic, usually epoxy resin, is used to impregnate 'blue john' so that thinner-walled and more delicate objects can be fashioned from it. If heat is used to cure the resin, care must be taken that the color is not lost by process **A**, although it could be recovered by process **B** if the resin does not darken at the same time. **C**

Although colored impregnation and dyeing do not seem to have been reported, their use on massive fluorite would not be too surprising. **D**

Plastic impregnation is detected by careful examination; the heat and irradiation treatments are not identifiable. #

References

1. K. Nassau, *The Physics and Chemistry of Color: The Fifteen Causes of Colour,* Wiley, New York (1983)
2. K. Przibram and J. E. Caffyn, *Irradiation Colours and Luminescence,* pp. 176–224, Pergamon Press, London (1956)
3. F. H. Pough, The coloration of gemstones by electron bombardment, *Z. dt. Ges. Edelstkeink.,* **20,** 71 (1957)
4. F. H. Pough and T. H. Rogers, Experiments in X-ray irradiation of gem stones, *Amer. Min.,* **32,** 31 (1947)

Fluorspar, *see* fluorite

Fossilized Bone, Teeth, Tusks, Horn, etc., *see* ivory

Fossilized Wood, *see* chalcedony

Garnet, *see* rhodolite

Goshenite, *see* beryl

Granite

Fine-grained granite has been dyed to produce an attractive turquoise imitation (R. Crowningshield, unpublished observation).

Greened Amethyst, *see* quartz

Gypsum

Including: alabaster, onyx (in part), satin-spar (in part), and selenite.

Summary of Treatments

Heat:
A Modify appearance [s]
Other:
B Bleaching [u]
C Dyeing [s or u, f]
D Painting, waxing, etc. [s or u]

Gypsum, $CaSO_4.2H_2O$ is called selenite when in single-crystal form, alabaster and sometimes onyx when in massive form, and satin spar when it is fibrous in nature (there is also satin-spar marble). The extremely soft nature of these materials, the Mohs hardness being only 2, limits their use to ornamental carvings, trays, covered boxes, and occasionally to bead or other gemstone use.

Webster (p. 298) reports that carved alabaster objects are often immersed in cold water, the temperature of which is slowly raised to boiling and then again slowly cooled; this changes the appearance of the alabaster so that it looks more like the much harder white marble. Heating may 'open pores' or produce cracks which permit better penetration of dyes. **A**

Bleaching of alabaster is the subject of recipe No. 2034 in Dick's collection[1], using soap and hot water, whitening, pumice, and even dilute nitric acid. He also gives recipes for hardening, polishing, and staining imitation alabaster made of plaster-of-Paris, recipes Nos. 2020 to 2031. **B**

Dyeing and colored oiling have often been used to make alabaster more attractive and, for example, to distinguish the two colors in chess boards and in carved chess pieces. Carved objects are also often waxed and may be coated with paint when in statue form. **C** **D**

Reference

1. W. B. Dick, *Dick's Encyclopedia of Practical Receipts and Processes or How They Did It in the 1870s*, pp. 200–201, Funk and Wagnalls, New York (1974)

Hackmanite, *see* sodalite

Haüyne, Haüynite, *see* lapis lazuli

Heliodor, *see* beryl

Heliotrope, *see* chalcedony

Hiddenite, *see* spodumene

Hornbill Ivory, *see* ivory

Howlite

Massive howlite, $Ca_2B_5Si_9(OH)_5$ is a soft, white, sometimes black-banded material occasionally used for beads and carving. It has been dyed various colors and may also be waxed. An elaborate carving was dyed blue to produce a superficial resemblance to turquoise; a test used to identify it was to etch the surface with acid and observe melting of the howlite with a hot needle[1].

Reference

1. R. Crowningshield, Dyed howlite, *Gems Gemol.*, **13,** 58 (1969)

Hyacinth, *see* zircon

Hydrophane, *see* opal

Indicolite, Indigolite, *see* tourmaline

Iolite

The deep blue color of some iolite, the gem form of cordierite, may have the same iron–titanium charge-transfer origin as does the blue in sapphire. The heat treatments used to modify the blue color of corundum could not be applied here, however, because of the rather low melting point.

Iris Quartz, *see* quartz

Ivory

Including: animal bone, burinut, coco de mer, corozo nut, deer horn, doom (doum) palm nut, elephant tusks and teeth, fossil bone, teeth, and tusks, hippopotamus teeth, hornbill ivory, hog teeth, ivory palm nut, narwhal tusks, rhinoceros horn, sea palm nut, sperm whale teeth, vegetable ivory, walrus tusks, etc.

Summary of Treatments

Heat:
- **A** Darken, produce cracking [s]

Irradiation:
- **B** Darken [s]

Other:
- **C** Bleaching [u]
- **D** Dyeing [s or u, f]
- **E** Painting [s or u, f]
- **F** Softening and shaping [s]

Ivory is, in the strict sense, the material that constitutes the tusks of the elephant. It is composed of about 65 percent of hydroxyapatite, $Ca_5(PO_4)_3OH$, and 35 percent organic matter, predominantly collagen (Webster, pp. 584–598). Extended works on elephant ivory, incorporating discussions of enhancements, include the rather old volumes by Maskell[1] and Kunz[2], as well as a 16-part serial publication by Webster[3].

The tusks, teeth and horns of many other animals, some of which are listed above, are used as ivory or ivory substitutes, as described by Webster. There are also a series of vegetable ivories, including the nuts of the Polynesian ivory-nut palm, the South American nut palm (corozo nut), the doom or doum palm (doom or doum nut), the sea palm (coco de mer), and the talipot palm (burinut); the vegetable ivories have been summarized by Brown[4]. The discussion that follows applies mainly to elephant ivory, but the same processes can be expected to work quite analogously with almost all of the other products listed.

Gentle heat has been used to give ivory an old, darkened appearance. This process has the tendency to produce cracking, which can be encouraged by dipping the ivory into hot water and then exposing it to the heat of a fire, which simulates the appearance of great age according to Kunz[2], p. 260. It should be noted that some ivories darken over centuries to a deep walnut, even an ebony color, while others remain very light in color[1]. These darkening processes involve the organic matter present in ivory, which can also be darkened by irradiation. **A** **B**

Bleaching ivory is usually carried out by applying hydrogen peroxide[5] or some chlorine-containing compound such as laundry bleach[2,6], or by exposing the ivory to sunlight behind a sheet of glass (Webster, p. 593). Other chemical processes are given by Kunz[2], pp. 262–263, and by Dick[6], recipe Nos. 1997–1999. Note that the 'sodium sulphate' of Kunz is obviously a misspelling of sodium sulphite as given correctly in Dick's recipe No. 1999. Peroxide bleaching has also been reported for the burinut[4]. **C**

Ivory has been dyed or stained for a variety of reasons. To produce an appearance of age, Maskell[1] mentions the application of oil followed by exposure to the sun, while Kunz[2] describes the use of tobacco smoke, tea, moist hay, or a bath of ochre. For artistic coloring, Kunz[2] mentions a wide variety of natural or synthetic dyestuffs; Dick[6] gives no less than 12 recipes (Nos. 1982–1993) for just about any color. Staining was also used on billiard balls in the days when these were made of ivory, before the advent of plastics. A walrus-ivory bead necklace has been stained a jade green, probably by a copper salt[7]. Heating ivory (or bone, etc.) and staining it with copper salts produces a blue odontolite imitation[8]. The red color present in hornbill ivory has also been applied artificially[9]. **D**

Painting on ivory for artistic purposes has been employed throughout the ages, particularly for the painting of miniatures. Kunz[2] describes the preparation of ivory with pumice powder to obtain an oil-free surface for good paint adherence. **E**

When treated with phosphoric acid (or nitric acid) ivory becomes soft and flexible; large sheets can then be prepared by peeling the tusk on a lathe according to Webster (p. 592), Kunz[2] (p. 263), and Dick[6] (recipe No. 1995). Hardening occurs by itself in air (but water is said to produce reversion to the softened state) or can be expedited by the application of 'dry decrepitated salt' according to Dick[6], recipe No. 1996. Maskell[1] discusses a whole series of ancient chemical treatments for softening, but states that after softening the ivory cannot be restored to the original hardness; he does not appear to have known of the phosphoric or nitric acid treatments. **F**

The detection of bleached or artificially darkened ivory is not usually possible. Materials stained unnatural colors, such as with the copper salts mentioned above, are usually readily recognized once ivory has been identified as present by conventional gemological testing. The absorption spectrum of odontolite is characteristic[10] and quite different from that produced by the copper stain. **#**

References

1. A. Maskell, *Ivories,* Methuen, London (1906); reprinted, C. E. Tuttle, Rutland, VT (1966) (with different pagination)
2. G. F. Kunz, *Ivory and the Elephant,* Doubleday, New York (1916)
3. R. Webster, Ivory – its varieties and imitations, *Gemmologist,* from **13,** 45 (Jun, 1944) to **14,** 2 (Aug, 1945)
4. G. Brown, The burinut, *Austral. Gemm.,* **13,** 134 (1978)
5. A. Singhee, Ivory carving in India, *Gem World,* **3,** 23 (1976)
6. W. B. Dick, *Dick's Encyclopedia of Practical Receipts and Processes or How They Did It in the 1870s,* pp. 198–200, Funk and Wagnalls, New York (1974)
7. E. A. Jobbins and P. M. Statham, Stained walrus ivory, *J. Gemm.,* **14,** 288 (1975)
8. H. Lee and R. Webster, Imitation and treated turquoise, *J. Gemm.,* **7,** 249 (1960)
9. R. T. Liddicoat, Jr, Hornbill ivory, *Gems Gemol.,* **13,** 98 (1969)
10. R. Crowningshield, Odontolite, *Gems Gemol.,* **12,** 21 (1966)

Ivory Palm Nut, *see* ivory

Jacinth, *see* zircon

Jade Group

Including: chloromelanite, jadeite, nephrite.

Summary of Treatments

Heat:
A Lightening green color, darkening yellow and brown, artificial ageing [s]
Irradiation:
B Darken [s]
Other:
C Dyeing [s or u, f]
D Colorless impregnation [s or u]
E Foil, backing, etc. [s or u]
F Composites [s or u, f]

The designation jade includes two distinct substances: jadeite, $NaAlSi_2O_6$ (which includes the iron-rich chloromelanite variety), and nephrite, $Ca_2(Mg,Fe)_5Si_8O_{22}(OH)_2$. The distinction between these is most easily made by the refractive index (jadeite is about 1.66, nephrite about 1.62) and the density (jadeite 3.3–3.5, nephrite 2.9–3.1). Both types consist of masses of interlocking fibrous crystals which give them their extreme toughness, even though the hardness is only 6.5 to 7 on the Mohs scale. Colors vary from colorless via many shades of green to black, but also include yellow, brown, lavender, and so on. An excellent overview is given by Webster (pp. 255–266); most books devoted to jade deal predominantly with its artistic and historical aspects. The treatments here described apply equally to both jadeite and nephrite, except when indicated.

A

Heating lightens the green color of jade, as has sometimes been noted during jewelry repairing[1]. Heating jade containing yellow-to-brown iron inclusions, as in the rind, may change these colors to browns and reds. Heating to ‘white heat’ and quenching in a concentrated acid, surely a very hazardous process, has been used to artificially age jade[2]. Heating, of course, destroys the color of dyed jade, such as the lavender type, even at the relatively low temperatures of 220–400 °C; other lavender jade (untreated?) retains its color to 750 °C, but not to 875 °C[3] or to over 1000 °C[4].

B

Irradiation generally produces only brown or blotchy colors which fade in sunlight or on heating to 100 °C, although a trace of a purple was noted in one specimen[3,5]. The possibility of a special (unknown) irradiation treatment which has been suggested as the origin for the abundance of ‘natural’ lavender stones seen in recent years has not been confirmed or eliminated at the time of writing.

C

Both jadeite and nephrite are dyed green or lavender colors, with dyed jadeite[6] being much more commonly seen because it tends to be more porous than nephrite[7], is more translucent and also is more frequently available in white or in pale colors. The irregular application (or absorption) of the color produces a more realistic appearance[8,9]. Green was the most commonly seen color until relatively recently, when quantities of a violet, lilac, or lavender color have appeared[4,10].

The dyeing process as used in Hong Kong in 1958 has been described by Ehrmann[11,12]: the warmed stones are merely dipped into an alcohol solution of aniline dyes, such as a mixture of No. 62 metanil yellow YK60 (or metanil yellow No. 1955) and acid sky-blue BS (or alizarin blue GRL150). As might be expected, many such dyes are not very light-stable and may fade in a matter of a few weeks[6]; other dyes may be more stable. Natural dyes, such as cranberry juice, have been mentioned and T. Talpey (unpublished observation) reports that even a solution of gentian violet (a mixed dye containing crystal violet that is used in medicine) can be used; this is, however, bleached on contact with a hydrogen peroxide solution.

D

Paraffin wax is often used as a final processing step in the tumbling of jade to hide small fractures close to the surface[13]. It is also used on cabochons and carved objects, where it can produce problems for the setter who may break the jade because he or she is not aware of the presence of flaws.

E

Foils have not been used much in recent years, but a green-colored backing that greatly enhanced the value of a semi-transparent jadeite ring-stone has been described[14]. A concave mirror with a small opening is frequently a part of the setting used for jade in Hong Kong and produces a significant improvement in appearance (R. Crowningshield, unpublished observation).

F

Composite stones have occasionally been used, the most interesting type being made of a hollowed-out, fine quality, translucent, white imperial jade which is placed over a second piece of jade shaped to fit into the hollow together with a green gelatinous substance, the whole being sealed with a third piece of jade cemented onto the back[11]; this type of triplet is illustrated in *Figure 5.7*. A more conventional triplet containing a green cement layer has also been reported[15].

#

The identification of the material as jadeite or nephrite, rather than one of the many types of jade imitations described by Webster, follows conventional gemological-testing procedures. The presence of dyed material is sometimes revealed under the microscope, particularly with immersion if the material is translucent, when localization of the color may become obvious. The spectroscope is of considerable help in distinguishing the chromium and iron lines seen in green jades (the strongest jadeite line is at 437 nm) from the broad absorptions seen in the red part of the spectrum in dyed and composite stones. Note, however, that very dark, green jade with a high chromium content may also show a broad absorption band. Close examination of the front and back surfaces (if exposed) and focusing into the stone will reveal the presence of a composite stone. A recent detailed study of the lavender jadeites by Koivula[4] indicates that a strong orange fluorescence under long-wave ultraviolet is observed in some dyed jades, but is absent in others; here too the concentration of dye in surface cracks can usually be detected. Both heat and strong acids destroy the dyed color, but neither is recommended for routine gemological testing!

The identification of artificially aged jade compared to long-buried, 'grave-old' jade presents problems which have not yet been solved[16].

References

1. C. F. Fryer (Ed.), Jadeite, dangers of heating during jewelry repair, *Gems Gemol.*, **18**, 103 (1982)
2. J. Sinkankas, Some comments on the artificial alteration of jade by burning, *Lap. J.*, **12**, 676 (1958)
3. K. Nassau, The effect of gamma rays on tourmaline, greenish-yellow quartz, pearls, kunzite, and jade, *Lap. J.*, **28**, 1064 (1974)
4. J. I. Koivula, Some observations on the treatment of lavender jadeite, *Gems Gemol.*, **28**, 32 (1982)
5. R. Crowningshield, Commercial implications of gamma radiation on gem materials, *Z. dt. Gemmol. Ges.*, **25**, 95 (1974)
6. R. Crowningshield, Faded dyed jadeite, *Gems Gemol.*, **11**, 100 (1963–1964)
7. R. Crowningshield, Dyed nephrite jade, *Gems Gemol.*, **11**, 363 (1965–1966)
8. R. T. Liddicoat, Jr, Unevenly dyed jadeite, *Gems Gemol.*, **14**, 350 (1974)
9. R. Crowningshield, Dyed jadeite, *Gems Gemol.*, **12**, 245 (1967–1968)
10. R. T. Liddicoat, Jr, Dyed light-violet jade, *Gems Gemol.*, **13**, 323 (1971)
11. M. L. Ehrmann, A secret (and expensive) process!, *Gemmologist*, **28**, 38 (1959)
12. M. L. Ehrmann, How to color jadeite, *Lap. J.*, **12**, 646 (1958)
13. R. Crowningshield, Paraffin . . . its pros and cons, *Gems Gemol.*, **14**, 84 (1972)
14. R. T. Liddicoat, Jr, Contrived color in jadeite, *Gems Gemol.*, **11**, 314 (1965)
15. R. T. Liddicoat, Jr, New type of jadeite triplet, *Gems Gemol.*, **11**, 369 (1965–1966)
16. R. Crowningshield, An age-old problem, *Gems Gemol.*, **14**, 83 (1972)

Jargon, Jargoon, *see* zircon

Jasper, *see* chalcedony

Kauri Gum, *see* amber

Kunzite, *see* spodumene

Kyanite

Pale blue kyanite (also known as cyanite or disthene) is reported to lose its color on heating to 1200 °C[1].

Reference

1. G. Smith, E. R. Vance, Z. Hasan, A. Edgar, and W. A. Runciman, A charge transfer mechanism for rose quartz, *Phys. Status Solidi*, **A46**, K135–K140 (1978)

Labradorite, *see* feldspar

Lapis Lazuli

Including: haüynite (haüyne), lazurite, noselite (nosean), and sodalite.

Lapis lazuli is a composite aggregate that contains grains of several related substances, all having the general formula $(Na,Ca)_8(Al,Si)_{12}O_{24}X$, where $X = SO_4$ for haüynite and noselite, (S,SO_4) for lazurite, and Cl_2 for sodalite; also present may be calcite, diopside, and pyrite inclusions.

Heating to red heat has been reported[1] to intensify the color of some pale blue lapis lazuli to yield a fine, dark blue, but this may produce an unattractive greenish blue in some specimens. Sodalite, however, is reported to lose its color on heating and regain it on irradiation[2], indicating the probable presence of a color center.

Dyeing, often using aniline dyes, is frequently practiced to improve the color of lapis lazuli, although even almost colorless material has been so treated[3]. Coloring for the disguising of white calcite inclusions is also done. Localization of the color in cracks can usually be seen, and a cotton swab soaked in acetone or nailpolish remover usually reveals the dye. This may not work, however, if a colorless wax has been used as a final step, in which case it must first be rubbed off[4, 5]; the wax is revealed by a hot-point examination. The colorless wax is apparently intended to hide cracks, improve the surface polish, as well as to prevent the dye from rubbing off onto clothes or skin.

The use of colored waxes (including blue shoe paste!) has also been reported[6].

References

1. M. Bauer, *Precious Stones,* Vol 2, p. 438, Dover Publications, New York (1968)
2. K. Przibram and J. E. Caffyn, *Irradiation Colours and Luminescence,* p. 250, Pergamon Press, London (1956)
3. R. Crowningshield, Dyed lapis lazuli, *Gems Gemol.,* **12,** 180 (1967)
4. R. Crowningshield, Lapis-lazuli mystery solved?, *Gems Gemol.,* **12,** 278 (1968)
5. C. F. Fryer, (Ed.), Dyed and wax-treated lapis lazuli, *Gems Gemol.,* **17,** 103 (1981)
6. R. Crowningshield, Notes on lapis-lazuli, *Gems Gemol.,* **11,** 337 (1965)

Lazurite, *see* lapis lazuli

Liddicoatite, *see* tourmaline

Limestone, *see* marble

Magnesite, *see* marble

Malachite

Malachite often takes a rather poor polish and paraffin wax is used to improve the appearance and hide small cracks; epoxy resins could be similarly used.

Marble Group

Including: bowenite, calcite, dolomite, limestone, magnesite, meerschaum, onyx marble, pyrophyllite, rhodochrosite, satin-spar marble, sepiolite, serpentine, smithsonite, soapstone, steatite, talc, travertine, verd-antique, and verdite.

Summary of Treatments

Irradiation:
A Producing color [s or u, f]
Heat:
B Lightening or removing color [s]
Other:
C Dyeing [s or u, f]
D Colorless impregnation [s or u]
E Colored impregnation [s or u, f]
F Painting and coating [s or u, f]

The materials collected together into this group are all soft substances, which are sometimes used for ornamental purposes when in massive form; many are carbonates and react with acids. Some occur only in solid-color form while others may be banded. The term marble by itself is herein used specifically for the calcite composition.

A Irradiation can produce a variety of color centers in calcite and marble, particularly yellow, blue, and lilac ones[1]; some of these may be stable to light, others may fade.

B Heating can bleach the colors in some calcite, marble, and in violet-colored dolomite; some of these colors can be restored by irradiation, while others cannot. Cracking to improve the penetration of dyes can also be induced by heating.

C Dyeing has been widely used to make these soft, easily shaped materials look like more valuable substances that are more difficult to carve. Thus serpentine has been dyed to look like jade (Webster, p. 354); calcite, dolomite and magnesite to imitate turquoise[2]; and marble to imitate coral[3], lapis lazuli[4], malachite, and so on.

Workers who prepare architectural marble have long stained their wares, and a nineteenth century book gives lengthy details[5]. Surface staining, deep staining, imitation veining and other decorations are all described. Dyes employed include vegetable substances such as saffron, litmus, cochineal, and indigo dissolved in alcohol, turpentine, urine, and so on, as well as water solutions of inorganic substances such as silver nitrate for a red, gold nitrate for violet-to-purple, arsenic sulfide for yellow, and copper salts for green. Some of these colorants can also be dissolved (or rather suspended) in wax, which penetrates into the marble, a process used in Greek and Roman times. The well-polished marble is heated to 'open the pores' and permit deep penetration, but not so much as to damage it or the dye; the degree of heat is correct when the solvent just bubbles. When several colors are used, a specific sequence of their application must be followed to avoid incompatibilities. Additional marble-dyeing recipes in Monton's *Secretos* of 1760 and Dick's collection of twelve recipes are discussed in Chapter 5, pp. 70–71.

D Colorless paraffin wax is routinely used to hide small cracks and improve the quality of the polish on marble and related materials, specifically on beads and carved objects; this also serves to protect the surface from accidental staining and also protects the colors when dyeing is used[5]. In Roman times marble statues were stained, painted, and waxed routinely; the concept of flawless, white statuary is a relatively recent one. **D, F**

Colored wax has been mentioned as an impregnant above. Colored plastic has also been used to change the appearance of marble; a necklace sold as turquoise has been reported[6]. Before 3000 BC blue-glazed steatite was used as an imitation of turquoise[7]. **E, F**

Identification of treatments in a material held in such little esteem is not performed in practice, but would follow conventional lines. #

References

1. K. Przibram and J. E. Caffyn, *Irradiation Colours and Luminescence,* pp. 236–239, Pergamon Press, London (1956)
2. T. Lind, K. Schmetzer, and H. Bank, The identification of turquoise by infrared spectroscopy and X-ray powder diffraction, *Gems Gemol.*, **19,** 164 (1983)
3. C. F. Fryer (Ed.), A coral substitute, dyed marble, *Gems Gemol.,* **17,** 226 (1981)
4. C. F. Fryer (Ed.), Lapis lazuli imitation, *Gems Gemol.*, **18,** 172 (1982)
5. Anon (M. L. Booth, translator from the French), *Marble Workers Manual,* pp. 31, 147–153, 222–227, Baird, Philadelphia (about 1870)
6. R. T. Liddicoat, Jr, Blue-dyed, plastic-treated marble beads, *Gems Gemol.*, **11**, 115 (1963–1964)
7. J. Ogden, *Jewellery of the Ancient World,* p. 138, Rizzoli, New York (1982)

Maxixe, *see* beryl

Meerschaum, *see* marble

Microcline, *see* feldspar

Moonstone, *see* feldspar

Morganite, *see* beryl

Morion, *see* quartz

Mother-of-pearl, *see* pearl

Narwhal Tusk, *see* ivory

Nephrite, *see* jade

Nosean, Noselite, *see* lapis lazuli

Novaculite, *see* chalcedony

Oligoclase, *see* feldspar

Onyx, *see* chalcedony and gypsum

Onyx Marble, *see* marble

Öolitic Opal, *see* opal

Opal

Including: cochalong, hyalite, hydrophane, öolitic opal, opalized bone, fossils, wood, potch, etc.

Summary of Treatments

A Restoration of water after heating [s]
B Dye [s or u, f]
C Colorless impregnation [s of u]
D Colored impregnation [s or u, f]
E Foil, coating, etc. [s or u]
F Composites [s or u]

Opal consists of tiny spheres of silica SiO_2, containing some water, cemented together with more silica, also containing some water; the total water content is usually several percent. The compositions of the spheres and the cementing material are slightly different, so that there is a small difference in refractive index between them. As a result, light can be scattered from the spheres (or from the spaces between the spheres) in such a manner as to produce diffraction. Precious opal, the subject of this section, results when the spheres are packed together in a very regular fashion (cubic or hexagonal close-packed) so that there is formed a three-dimensional diffraction grating, which can diffract the different colors in white light to form the the play-of-color in opal. The spheres also must all have just about the same size, typically 200–300 nm in diameter. This insight resulted from electron microscope studies by Sanders and co-workers[1–3] at the Commonwealth Scientific and Industrial Research Organization in Melbourne, Australia.

If the spheres are all almost exactly the same size and very well ordered, then only a little white light is scattered and the play-of-color originates in an almost transparent material. With some variation in the sphere size, there is some scattered white light, called 'opalescence', and ordinary white opal results. Under special conditions, such as occurred in parts of Australia, the body color can be very dark, leading to the most desired black opal. Opals with other body colors are also known. Oolitic (or öolitic) opal is a rare variety of opal containing small spheres in structures which suggest that organic material or coral has been replaced (Webster, p. 240).

Hydrophane opal is the name used for a porous material which does not show color until it is immersed in water or some other fluid. It appears that the spaces between the spheres are incompletely filled in such a way that a large difference in the refractive index causes so much irregular light-scattering that the result is merely an opaque white color. When the water penetrates into the spaces between the spheres so that the desired, small refractive-index difference results, then the stone becomes translucent with play-of-color; this is again lost when the water evaporates.

A Heating opal drives off the water; this makes the refractive index uniform throughout the opal and results in loss of color and/or produces white opacity. Merely soaking the burned opal in water does not restore the color. When the opal collection of the Hungarian National Museum in Budapest was damaged by fire, a process was developed for restoration[4]. This involves exposing the stones to a vacuum of at least 700 mmHg, which removes air from the pores and permits water to penetrate deeply, with recovery of the play-of-color. Heating the restored opals to 130 °C does not affect the color; the color is lost at 700–800 °C and can again be restored by the same process. It is even conceivable that this treatment might work on some naturally occurring materials which do not show the play-of-color. **A**

A wide variety of processes have been used to dye opals; this is possible because of their slightly porous nature. Most frequently seen is black-dyed opal, produced by the same sugar–acid process described for chalcedony above and in Chapter 5, pp. 66–67. This process is particularly suitable for Andamooka, Australia matrix opal; this contains significant amounts of kaolinite clay which scatters white light and diminishes the play-of-color. An excess of scattered white light is then absorbed by the black carbon deposited within the pores, which intensifies the color and results in a product resembling good-quality black opal. The process[5–7] was apparently developed about 1960 and begins with the preforming, washing and drying of the stones at a little below 100 °C. They are then placed into a hot sugar solution (perhaps 2 cups of sugar and 3 cups of distilled water, or honey, corn syrup, etc.) that has been boiled, and left to soak in the hot solution for several days. After a slow cool and a brief rinse, and/or a quick wipe, the stones are placed into concentrated sulfuric acid, which is heated close to 100 °C for one or two days and then cooled slowly. Next comes thorough washing, followed by a brief rinse in a bicarbonate of soda solution and another wash. The stones are now ready for polishing. The depth of penetration depends on the porosity of the opal and on the length of the soaking. Even if there is not much penetration into the opal layer itself, the blackening of the matrix layers alone can produce a tremendous improvement in the appearance. **B**

Other substances mentioned for dyeing opal black include solutions of aniline dyes[8] and of silver nitrate. In at least one instance[9] only the ironstone matrix on the back of an opal cab was dyed for uniformity of the cabs, without, however, affecting the appearance from the front!

This process has also been applied to öolitic opal, where the dyeing may blacken either the material inside the spheres or the material between the spheres[10, 11]. Opal matrix or 'potch' itself is dyed by the same process to serve as backing in the opal composites described below.

An interesting 'dyeing' is used for the so-called 'smoked' opal. This treatment extends back at least to 1947[12] and the product is seen occasionally; it is applied to extremely porous hydrophane opals, particularly those from Jalisco, Mexico, to produce material looking just like Australian black opal. Several variants of the process have been described. The preformed opals may be wrapped in layers of newspaper and covered with aluminum foil[13], in brown wrapping paper[14], or soaked in old crankcase oil and wrapped in brown paper[15]; the whole is then strongly heated to carbonize the paper. Alternatively, the stones are buried in fertilizer (manure) in an earthenware jar and baked over charcoal[14]. Although usually called smoke-impregnation, it is obvious that considerable oily and tarry matter is distilled out of the newspaper or fertilizer, so that the product is more akin to a colored oiling, even though smoke-like carbon particles are also present, as in the sugar–acid process, with which this process could easily be confused. The result is a play-of-color in an oil-filled opal which still remains quite porous and can lose its color on being made wet (probably caused from excessive light scattering by the mixture of oil and water) but which regains its color on drying[15, 16]. **B**

A colorless impregnation can be used on hydrophanes or on worthless-appearing potch opal to produce an excellent play-of-color. Oils (and waxes) of many types[8] are frequently used, not only to improve the play-of-color but also to hide cracks to which some opals can be prone. The oil frequently dries out or seeps out relatively rapidly. The same is true of glycerin used to hide cracks. Very deceptive can be a plastic impregnation which is stable and can result in the drastic transformation illustrated in *Plate XXV*, producing the change from a chalk-like potch to the brilliantly colored appearance of quality opal. Although details have never been revealed, the process undoubtedly involves drying followed by a vacuum impregnation. Both plastic as well as silica-based (silane) polymers appear to have been used, according to Manson[17, 18], where a detailed description may be found. **C**

The smoke treatment discussed above is in fact one form of colored impregnation. Black plastic has also been used to impregnate porous opal[19]. **D**

Reflective foils, peacock feathers, and mother-of-pearl are some of the substances which have been placed behind translucent opal to 'improve' the play-of-color[20]. Black paint has also been used on the back of cabochons to absorb light and improve the intensity of the play-of-color. Coatings could be used to protect the surface of opal; an acrylic layer is applied to the all-plastic opal imitation to protect its extremely soft polystyrene body[21], but is itself still relatively soft. **E**

Composite structures have been widely used for opal to protect it with a harder top, to utilize thin seams, as well as to provide a dark backing for an improved play-of-color[22]. In doublets a layer of opal may be cemented to a base of black glass, black obsidian, black onyx (dyed chalcedony), dyed opal matrix (potch) and so on. Triplets may consist of a quartz or glass top, a thin opal center, and a black glass, onyx, or matrix back; alternatively, almost any back can be attached with a black cement. Synthetic sapphire has even been used as the crown in doublets or triplets to provide a truly scratch-resistant top. **F**

Other substances, such as mother-of-pearl and even dyed fishskin in a 'Schnapperskin triplet', have been used to simulate the play-of-color (Webster, p. 466). Pieces of opal are sometimes suspended in a fluid such as water and/or glycerin in a glass or plastic container to produce 'floating opal'. The misnomer 'reconstructed opal' (or 'fragmented opal-doublet') has been suggested for a stone made by cementing small chips of opal together with a dark epoxy resin or the like[23]. Triplets[24] have even been made using Gilson synthetic[25] opal!

Identification begins with the distinction of natural opal from the synthetic along conventional lines. Careful microscopic examination reveals the presence of carbon particles looking like grains of pepper in the dyed material and the concentration of dye in cracks. Dyeing observed on the back of a cabochon does not necessarily extend to the front[12]. Smoke-treated opal is very porous (it sticks to the tongue), has a low refractive index (1.38 to 1.39) and specific gravity, and loses its play-of-color when wet[15, 16]; the color returns again when dry. #

Colorless oils, waxes, or plastic may present problems[26], but can usually be detected by careful observation. A low refractive index and, in a loose stone, a low specific gravity immediately suggest the presence of an organic impregnant or a smoke-treated stone. Careful application of a hot point to an inconspicuous part of the stone can reveal oil, wax, or plastic. A solvent may remove a little of the oil and reveal cracks in a 'slightly destructive' test. Similarly, the application of a needle produces a scratch in plastic quite different from that in untreated opal. Brazilian plastic-impregnated opal may contain tiny, opaque inclusions of the nickel–iron sulfide bravoite[18].

Doublets or triplets should present no problem in loose stones; they are usually revealed in a mounting when the front and back surfaces are carefully compared. Sometimes it is possible to see gas bubbles in the cement layer from the top. With a completely enclosed setting it may be necessary to unmount the stone for a positive identification. Certain natural opals can look deceptively like doublets. In doublets and triplets it is also necessary to look out for the possibility of synthetic, rather than natural, opal[24].

References

1. P. J. Darragh, A. J. Gaskin, B. C. Terrel, and J. V. Sanders, Origin of precious opal, *Nature,* **209,** 13 (1966)
2. J. B. Jones, J. V. Sanders, and E. Segnit, Structure of opal, *Nature,* **204,** 990 (1964)
3. J. V. Sanders, Colour of precious opal, *Nature,* **204,** 1151 (1964)
4. E. Hunek, Regenerated precious opal, *Gemmologist,* **30,** 101 (1961)
5. W. H. Walker, A new type of simulated heat-treated 'black opal', *Lap. J.,* **17,** 655 (1963)
6. W. A. Rose, Treating matrix opal, *Gems Gemol.,* **14,** 306 (1974)
7. R. A. Ball and N. Clayton, Opal references and abstracts, *Austral. Gemm.,* **12,** 181 (1975)
8. Anon, Chemical coloring of precious stones, *Jewelers' Circ.-Keyst.* (Feb 2, 1927)
9. R. Crowningshield, Stained-black opal, *Gems and Gemol.,* **13,** 249 (1970–1971)
10. R. Crowningshield, Black-dyed oolitic opal, *Gems Gemol.,* **13,** 351 (1971)
11. C. F. Fryer (Ed.), Oolitic opal, *Gems Gemol.,* **18,** 104 (1982)
12. D. E. Mayers, Mexican black opal, *Gems Gemol.,* **5,** 475 (1947)
13. E. Reed, An opal tip, *American Opal Society Newsletter* (Dec 6, 1972)
14. R. T. Liddicoat, Jr, Notes on recent synthetics and the blackening of natural opals, *J. Gemm.,* **12,** 309 (1971)
15. A. M. Waddington, Letter to the editor, *Austral. Gemm.,* **12,** 254 (1975)
16. R. Crowningshield, Loss of color in opal, *Gems Gemol.,* **12,** 179 (1967)
17. D. V. Manson, Plastic impregnated gem opal, *Gems Gemol.,* **16,** 49 (1978)
18. D. V. Manson, Plastic impregnated opal and plastic opal, *Gems Gemol.,* **16,** 219 (1979)
19. R. T. Liddicoat, Jr, Unusual opals, *Gems Gemol.,* **13,** 148 (1970)
20. M. Bauer, *Precious Stones,* Vol 2, p. 377, Dover Publications, New York (1968)
21. N. Horiuchi, New synthetic opal made of plastic, *Austral. Gemm.,* **14,** 213 (1982)
22. E. Gübelin, New fakes to simulate black opal, *Gemmologist,* **28,** 141 (1959)
23. F. Leechman, *The Opal Book,* p. 233, Lansdowne Press, Sydney (1961)
24. G. Lenzen, Tripletten aus synthetischem Opal, *Z. dt. Gemmol. Ges.,* **26,** 85 (1977)
25. K. Nassau, *Gems Made by Man,* pp. 257–267, Chilton, Radnor, PA (1980)
26. R. K. Mitchell, Oiled opals, *J. Gemm.,* **18,** 339 (1982)

Opalized Bone, Fossil, Wood, etc., *see* opal

Operculum, *see* pearl

Orthoclase, *see* feldspar

Padparadscha, *see* corundum

Pearl

Including: abalone, conch pearl, mother-of-pearl, operculum, and other shell materials.

Summary of Treatments

A Irradiation to darken [s or u]
B Bleaching [s or u]
C Restoring luster or color by peeling; healing cracks; etc. [s or u]
D Dyeing [s or u, f]
E Coating, composites, etc. [s or u]

Pearls are the product of several mollusks, most frequently of the salt-water species now known as *Pinctada.* Other salt-water species include *Strombus* and *Haliotidae*, the abalones. Fresh-water mollusks include members of the *Unio* and *Quadrulo* species. The pearls may be produced naturally or may be cultured by the insertion of mother-of-pearl beads, half beads, or pieces of mantle tissue into the mollusk.

The attractiveness of the pearl derives from the slightly iridescent reflection (often called 'orient') from the surface layer or nacre, produced by tiny platelets of aragonite, $CaCO_3$, held together by about 4 percent of conchiolin; the latter may give a body color to the pearl. Much detail is given by Webster (pp. 501–561). There are many books on pearls, with a particularly good, if old, one being that by Kunz and Stevenson[1].

A Irradiation with gamma rays or other energetic radiation produces a darkening; some species are more susceptible than others. This so-called 'black' color is, in fact, a gray or bluish gray[2], and is shown in *Plate I.* This process is used on those greenish and other off-color cultured pearls that do not respond in a satisfactory manner to bleaching. Fresh-water Lake Biwa Pearls were found to turn a silver–gray color on neutron irradiation[3].

B Bleaching has been widely practiced on both natural and cultured pearls[4] since at least 1924[5]. The process employs typically '10 volume' hydrogen peroxide, used for a few days to two weeks at about 40 °C, often with exposure to sunlight or ultraviolet. When dark patches occur below the surface layer of the pearl, holes may be drilled from the stringing hole toward the patches to permit internal access to the bleaching solution[4]. Much stronger reagents, such as chlorine, have been used, but may cause the pearl to become brittle and friable[1]. Even peroxide can produce a chalky surface (Webster, p. 538), presumably if applied inexpertly.

In the 1920 book by Rosenthal[6], it is claimed that pearls need to 'dry' in air for two to three years, during which time they become whiter. A chemist was reported to have found a way of 'washing out the sea water' and so lightening the pearls in two weeks. It is not difficult to see a 'cover story' for peroxide bleaching in this fanciful tale.

C The luster of pearls may become dulled and they may be stained and yellowed by contact with perfume, oils, and the acid skin-secretion of some individuals.

Many techniques have been proposed over the centuries to revitalize pearls, including feeding them to fowl, as first reported about 400 AD in *Papyrus Holmiensis*, as described in Chapter 2, p. 9; it was undoubtedly the acidic digestive fluids which lightly etched the surface. Baking in a loaf of bread was mentioned by Kluge in 1860, as also described in Chapter 2, p. 20. **C**

The process of 'peeling' or 'skinning' a pearl, namely carefully splitting off the surface layer with a blunt tool in the hope of finding a better surface beneath, has been described at length in many of the older pearl books[1,6]. This is obviously a task for a real expert. Sometimes each peeling leads to another one, until no pearl is left! In one instance a natural black pearl, which had faded in sunlight, was successfully peeled to reveal a new black surface[7]. Peeling appears to go back to Pliny's time (about 55 AD), for he says[8]: 'Pearl . . . is formed with a skin of many thicknesses, . . . consequently experts subject them to a cleaving process'.

In Vertrees' book[9] there is a fanciful tale that: 'Pearls worn by invalids for a long time sometimes assume a dull color and are called 'sick pearls' '. The proposed quaint solution is: '. . . to let servants wear them. This often improves the luster'.

Minor surface blemishes can be removed by a mild abrasion. It is reported that small surface cracks can be 'cured' by soaking pearls in warm olive oil; a dark brown color can be produced, however, with overheating at a temperature as low as 150 °C[10] (Webster, p. 520).

Pearls are dyed a variety of colors by several techniques. For pale tints (such as those permitted without disclosure by some guidelines – *see* Chapter 6) the pearls are usually first bleached with hydrogen peroxide[5]. For the frequently desired (in the US) 'rosée' or 'rosetint' color, a dilute oil or alcohol solution of the red dye eosin is often used (Webster, p. 538). Other dyes reported[10] include natural organic extracts, aniline dyes, as well as inorganic substances such as cold potassium permanganate solution for a brown[11] and a combination of alkali and cobalt salt, followed by ultraviolet light, for a pink[12]. Fading can occur, depending on the colorant used. **D**

The most frequently used dyeing process on pearls is silver staining to produce a non-fading, deep, rich black color. This employs silver nitrate dissolved in a dilute ammonia solution, followed by exposure to light or hydrogen sulfide gas[11] (Webster, pp. 522, 538).

Dyeing has even been employed on the mother-of-pearl beads used in the culturing process to produce a body-color modified by the underlying bead color; a cross-section of such a cultured pearl is shown in *Figure 7.11*. Selective area-dyeing has been achieved by drilling holes from the stringing hole toward the desired surface region and injecting the dye into the hole[4,13]. This process is called a center treatment to distinguish it from the more usual surface treatment.

Both colorless coatings of collodion or plastic, as well as colored coatings, have been used to modify the luster and color of pearls[6,13,14]. The only composite normally used is the Mabe pearl or cultured blister pearl. This is made from an incomplete pearl grown over a pellet attached to the shell of a mollusk. On being harvested the pellet is removed and the shell is waxed or cemented over a mother-of-pearl bead and backing[15]. **E**

The identification of pearls and pearl treatments is a task for the expert. It begins with the separation from pearl imitations and the distinction between **#**

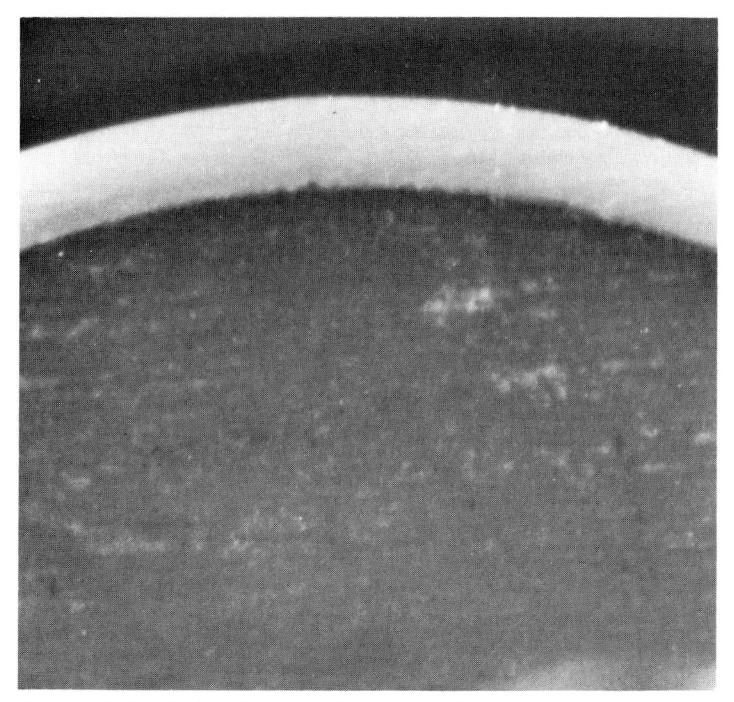

Figure 7.11 Section of a 'seed-dyed' pearl, showing the dark color of the seed and the light color of the cultured growth

natural and cultured pearls by standard gemological testing[15, 16]; the endoscope and X-ray radiography may have to be employed, and they also reveal the internal structure of Mabe pearls. Bleaching or pale-colored tinting are not necessarily recognized, but more intense color dyeing may show color concentration in cracks, in holes made for center treatment, in the stringing hole, and even on the string itself.

Only quite recently could it still be stated[15] that black pearls, if cultured, had definitely been colored, but this is no longer true since some limited production of cultured black pearls has been recently reported[17]. Natural black pearls colored by the silver process described above reveal their nature on radiography (the silver strongly absorbs X-rays), by X-ray fluorescence analysis (which identifies silver if present), by photography using infrared-sensitive film, by the lack of the normal red fluorescence under long-wave ultraviolet radiation, and so on[13, 15–18]. Black and other color dyes may be detected by rubbing off onto the string or by a white swab dipped in dilute nitric acid (1 part in 20 parts of water), which may remove a little of the color (and could destroy the orient if used carelessly). So-called 'black' irradiated pearls cannot be identified with certainty, but are in fact not black in color, but various shades of gray.

References

1. G. F. Kunz and C. H. Stevenson, *The Book of The Pearl,* Century, New York (1908)
2. K. Nassau, The effect of gamma rays on tourmaline, greenish-yellow quartz, pearls, kunzite, and jade, *Lap. J.*, **28,** 1064 (1974)

3. R. T. Liddicoat, Jr, Irradiated cultured pearls, *Gems Gemol.,* **12,** 153 (1967)
4. R. Crowningshield, Bleached and dyed cultured pearls, *Gems Gemol.,* **11,** 99 (1963–1964)
5. S. Shirai, *The Story of Pearls,* pp. 91–92, Japan Publications, Tokyo (1970)
6. L. Rosenthal, *The Kingdom of the Pearl,* pp. 32–50, Brentano's New York (1920)
7. E. W. Streeter, *Pearls and Pearling Life,* pp. 264–265, London (1886)
8. H. Rackham, *Pliny: Natural History,* Vol 3, Book 9, Ch 54, p. 237, Harvard University Press, Cambridge, MA (1940)
9. H. H. Vertrees, *Pearls and Pearling,* pp. 192–193, Fur News Pub, New York (1913)
10. H. Lee and R. Webster, Notes on discoloration of pearls, *Gems Gemol.,* **4,** 273 (1954)
11. X. Saller, Das Färben von Perlen, *Z. dt. Ges. Edelsteink.,* **25,** 12 (1958)
12. Anon, Coloured cultured, *Gemmologist,* **19,** 448 (1950)
13. L. B. Benson, Jr, Testing black pearls, *Gems Gemol.,* **10,** 53 (1960)
14. R. Crowningshield, Plastic-coated cultured pearls, *Gems Gemol.,* **12,** 116 (1966–1967)
15. B. W. Anderson, *Gem Testing,* 9th Edn, pp. 384–402, Butterworths, London (1980)
16. R. T. Liddicoat, Jr, *Handbook of Gem Identification,* 11th Edn, pp. 163–178, Gemological Institute of America, Santa Monica, CA (1981)
17. H. Komatsu and S. Akamatsu, Differentiation of black pearls, *Gems Gemol.,* **16,** 7 (1978)
18. R. Crowningshield, The spectroscopic recognition of natural black pearls, *Gems Gemol.,* **10,** 252 (1961–1962)

Perthite, *see* feldspar

Petrified Bone, Coral, Wood, etc., *see* chalcedony

Phenakite

Colorless material has been reported to turn a yellow–brown color on irradiation[1]. One could expect this change to be reversed by heating.

Reference

1. F. H. Pough and T. H. Rogers, Experiments in X-ray irradiation of gem stones, *Amer. Min.,* **32,** 31 (1947)

Plagioclase, *see* feldspar

Plasma, *see* chalcedony

Potch, *see* opal

Prase, *see* chalcedony

Prasiolite, *see* quartz

Pyrophyllite, *see* marble

Quartz

Including: amethyst, aventurine quartz, blue quartz, cat's-eye quartz, citrine, 'greened amethyst', greenish yellow quartz, iris quartz, morion, prasiolite, quartzite, rainbow quartz, rock crystal, rose quartz, smoky quartz, 'smoky topaz', tiger-eye, etc. (but generally excluding the cryptocrystalline chalcedony family discussed under chalcedony).

Summary of Treatments

Irradiation:

A Colorless and pale colors to smoky or greenish yellow [s, r]
B Yellow or green to amethyst [s or u, f?, r]
C Colorless to rose [s or u, f?, r]

Heat:

D Smoky to paler to greenish yellow to green, blue–green, or blue and on to colorless [s, r]
E Amethyst to yellow (citrine) or bicolor amethyst–citrine; all [s], some [r]
F Amethyst to colorless to green; all [s], some [r]
G Rose to lighter to colorless; all [s]; some [r]
H Blue to modified or colorless [s]
J 'Crackled' for iris quartz or dyeing, milky [s]
K Yellow or brown to red–brown or red [s]

Other:

L Bleaching and dyeing [s or u, f]
M Composites [u, f]
N Surface coating [u]
O Foils, mirrors, etc. [u]

Some of these treatments can also be applied to some of the cryptocrystalline varieties of quartz described under chalcedony; analogously, some of the treatments there described also apply to some of the crystalline forms discussed here.

It would be easy to fill a whole book with the description of the hundreds of studies that have been published over the years on the causes of the colors produced by impurities, irradiation, and heat in quartz. Since this is outside of the scope of this book, only a brief outline of a few of the well-established facts is included here, with enough references to lead the interested reader back into the detailed literature.

Quartz is the room-temperature alpha-form of crystalline SiO_2. In the absence of color-causing impurities, such as iron or titanium, and without color centers, quartz is colorless, sometimes called rock crystal. Iron in the Fe^{3+} ferric state can produce a yellow-to-brown-to-reddish color called citrine (or 'topaz' or 'smoky topaz', highly undesirable misnomers), or a green color called prasiolite if in the Fe^{2+} ferrous state. Titanium probably causes the color in pink rose-quartz, in combination with iron. Titanium can also produce blue quartz, asteriated quartz, or rutilated quartz, all containing needles of TiO_2.

The colors of smoky quartz and greenish yellow quartz, also called morion and 'smoky topaz' in part, are derived from a color center that involves an aluminum impurity, as described in Chapter 4, pp. 55–56. As also described there, a color center associated with Fe^{2+} or Fe^{3+} impurities gives the color to amethyst.

Inclusions of asbestos fibers, usually crocidolite in its original state in hawk-eye, but frequently highly altered to pseudocrocidolite, produce the cat's-eye and tiger-eye forms of quartz. Inclusions of mica and other colored minerals in single-crystal quartz or in granular, polycrystalline quartzite produce a variety of colors, usually designated aventurine quartz.

The best single source on quartz is Frondel's volume[1], but much of the insight on the causes of color came after its publication. The gemological aspects are summarized by Webster (pp. 199–211). Several of the color-causing mechanisms can occur simultaneously, thus further extending the range of colors observed in quartz. There is a lack of agreement on the definition of the term citrine; some would restrict it to naturally found yellow-to-orange quartz and use 'burned amethyst' or the like for the heated material. Herein, the usage of Webster, Frondel[1], Bauer[2], and most gemologists, is followed with citrine referring to an iron-caused yellow color, however produced.

The Smoky to Greenish Yellow to Colorless Changes

The irradiation of almost any colorless quartz produces a smoky color, which **A** can range from black through gray and brown to reddish, yellowish, and greenish tints. Studies over many years (with much confusion along the way) have established that the smoky color is derived from the broad 'A_3' absorption band at about 440 nm, which is usually accompanied by the broad 'A_1' and 'A_2' bands at about 690 and 500 nm, respectively[3,4]. These bands are shown in *Figure 7.12*. The

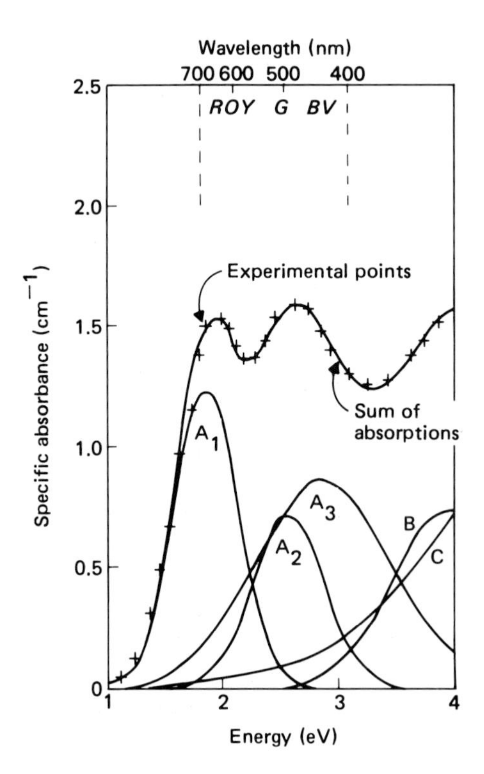

Figure 7.12 The absorption spectrum of a smoky quartz; the crosses are the experimental points and the solid line passing through them gives the sum of the underlying components. A_1 and A_2 together, but without A_3, give a blue color; A_3 gives the gray-to-black smoky color; B and C by themselves give a yellowish color. After K. Nassau and B. E. Prescott, reference 4

A

'A_3' band originates from a color center which consists of a hole (a missing electron) on one of the four oxygens next to an aluminum ion that has substituted for silicon, with the electron trapped at a nearby hydrogen or alkali ion, as described in Chapter 4, *Figure 4.7*, pp. 55–56. Models for the 'A_2' and 'A_1' centers have also been given[5], with 'A_2' involving an electron missing from two of the four oxygens next to the aluminum and 'A_1' involving an electron missing from three of the four oxygens. The role of hydrogen in accepting the displaced electrons is quite complicated[6].

Since essentially all quartz, including synthetic quartz[7], contains some aluminum, all quartz produces a more or less intense, smoky color on irradiation and this may be modified by the simultaneous presence of some amethyst, greenish yellow, or other colors. Essentially any form of irradiation (but not ultraviolet) produces the smoky color, which is stable to light. In synthetic quartz there is frequently observed an anomalous brownish–greenish color variation, derived from an anisotropy induced by growth on opposite sides of a rhombohedral seed[8]; this has sometimes been incorrectly called a form of twinning. The smoky color often occurs in bands or growth zones[1]. Irradiation has been used in a patented process[9] to select natural quartz with low aluminum content. Some natural and synthetic quartz turns a greenish yellow color on being irradiated, without any trace of smoky quartz[4].

D

Heating smoky quartz produces a lightening and subsequently a total loss of the smoky color. This process is time- and temperature-dependent; in one specimen it may require weeks at 180 °C, days at 200 °C, hours at 300 °C, and merely minutes at 400 °C[1]. For a large collection of different specimens heated for 2–4 h, the smoky color completely disappeared at temperatures ranging from a low of 140 °C to as high as 380 °C[4]. There is one unique exception to these temperatures. If electrolysis is used to produce smoky quartz, the electrons lost in forming the color centers are removed from the quartz; as a result this smoky color is now stable to over 1000 °C[10].

On being heated, much natural, smoky quartz or irradiated quartz passes through a greenish yellow color on its way to colorless[3,4]. This greenish yellow color, shown in *Plate IX*, is stable to light and has also been called honey quartz or even the much less desirable name, 'radiation-produced citrine'. No iron is contained in this material, the color deriving from a color center of unknown nature. Such a greenish color has been observed on many occasions[11] and is returned back to the smoky color on re-irradiation. It has been noted as occurring in what where probably naturally heated, smoky quartz crystals[12]. Greenish yellow quartz forms either directly during the irradiation of colorless quartz (very rarely without any smoky coloration) or by heating for 2–4 h at temperatures ranging from as low as 140 °C to as high as 280 °C; the color is similarly lost at from 140 °C to 380 °C[4].

On very rare occasions a blue color is seen on heating, as an intermediate step between the smoky and the colorless states[4]. This color is stable to light and consists of the 'A_1' and 'A_2' absorption bands by themselves, without any 'A_3'. The blue forms at 160–280 °C and this color is lost at 280–360 °C. Occasionally blue–green and green colors are observed, which are merely mixtures of the blue, brown, and yellow color centers.

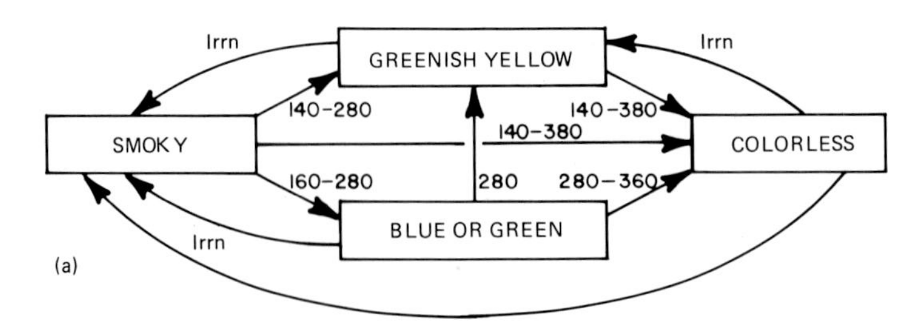

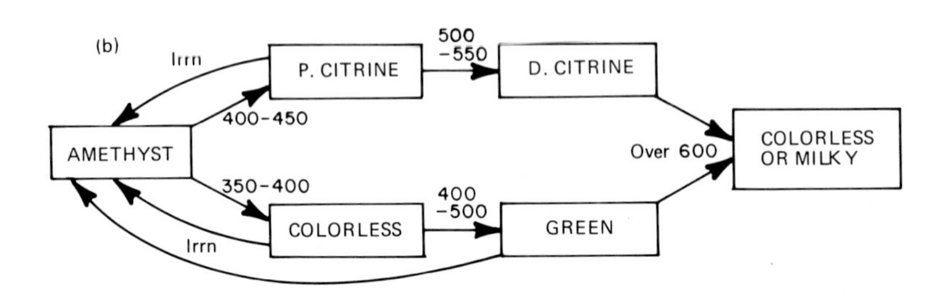

Figure 7.13 The heat and irradiation changes involving (a) smoky quartz and (b) amethyst. The temperatures shown are in °C and apply to heating times of several hours

The various color changes on irradiation and heating discussed in this section are summarized in *Figure 7.13(a)*. **A,D**

The Amethyst to Citrine or Prasiolite Changes

Irradiation of quartz containing iron can produce the bluish-to-reddish violet color of amethyst. Studies based on synthetic quartz[7, 13] have shown that both yellow Fe^{3+}-containing and green Fe^{2+}-containing quartz yield amethyst on irradiation by forming the same color center as described in Chapter 4, pp. 55–56. This, however, only happens if growth has occurred at certain growth orientations, such as in the *r* or *z* rhombohedron or, rarely, the *Z* basal-plane directions; it does not occur in growth in the prism directions[1, 7]. Natural amethyst almost always shows polysynthetic twinning on the Brazil Law, again with growth on the rhombohedron *r* and *z* faces[1]. At times, the irradiated quartz has a combination smoky–amethyst color. Careful heating to remove the smoky color by process **D** without initiating the loss of the amethyst color by processes **E** and **F** then improves the colors. **B**

Some amethyst is reported to fade on being exposed to bright light[14, 15] (Webster, p. 210). It is known that fading in light is accelerated by relatively mild heating and it is not clear whether amethyst fades in light without the presence of some heat.

Heating of amethyst produces a lightening of the color, as shown in *Plate XVII,* followed by one of two changes: if Fe^{3+} was present before irradiation, whether natural or induced by human action, there develops a yellow, usually called citrine but sometimes 'burned amethyst' (particularly in the German **E,F** literature[16]), while if Fe^{2+} was present, a green color develops, sometimes called prasiolite. These transformations are exactly the reverse of those described in process **B** above. **E,F**

Most of the citrine ('smoky topaz') seen in the trade is made by heating amethyst at about 450 °C[1, 15, 17]; this again is a time-dependent process. In practice, heating is begun at about 150 °C and the stones are removed and inspected at about 25 °C intervals; at each step those that have reached what is considered to be the optimum color are removed. At first, the amethyst color lightens, as in *Plate XVII,* and then the yellow appears. Occasionally, heating at a relatively low temperature has been stated to produce a colorless quartz[15], but this derives from the fact that only a very small amount of Fe^{3+}, giving a barely detectable yellow, is sufficient to form a reasonably dark, amethyst color on irradiation. Heating citrine or amethyst to slightly higher temperatures, in the 500–575 °C region, as given by Wild in *Table 3.1,* intensifies the yellow color, converting it into the orange, brown, or reddish color typical of heated iron-oxide compounds. Heating to yet higher temperatures may completely bleach the color, and produce a milky opalescence[1], as seen in *Plate XVII.* Those changes are illustrated in *Figure 7.13(b).* **E**

The conversion from amethyst into citrine often occurs at slightly different temperatures, even within a single specimen, so that careful heating to about 350–400 °C can sometimes produce[17, 18] a bicolor amethyst–citrine, also called 'ametrine'; a gemological examination of such material has been published[19]. Alternatively, heating to a somewhat higher temperature (400 or 450 °C in the cited work[17, 18]) converts all the amethyst into citrine, but only some may return to amethyst on irradiation, with the same end-result, as shown in *Plates VII* and *VIII.* This material can also be produced by the irradiation of synthetic iron-containing quartz, grown in different directions from a single seed; the *r* growth turns amethyst, but the *Z* growth does not[17, 18].

In a most unlikely behavior, a Brazilian amethyst, after being heat bleached, reverted by itself, over a seven-month period, into a pale violet color[11]. One suspects that this specimen had been inadvertently stored next to some radioactive mineral specimen!

Some amethyst turns green on being heated, and is called prasiolite at times. This appears to have been first discovered in amethyst from Montezuma, Minas Gerais, Brazil[20], although this change had been observed, but not recognized as significant, long ago[11]; green-turning material also occurs at other localities, such as Four Peaks, Arizona[21]; in Zimbabwe (Webster, p. 210); at Cearã, Brazil; and in a very interesting location in California, 33 km north of Reno, Nevada. In this last area there are found, over a 3 km distance, amethyst, citrine, and green quartz. It is believed that heat from a nearby rhyolite inclusion converted the amethyst closest to it into citrine or green quartz while the more distant amethyst was not affected[22]. **F**

The change from amethyst to green sometimes seems to go through a colorless state[15], as indicated in *Figure 7.13(b),* resulting from an exact balance of the complementary green and purple colors during the conversion. The green is produced at about 400 to 500 °C and stronger heating bleaches the green, producing a milky appearance, as with citrine. Webster (p. 210) questions the stability of some green quartz to light, but no specific studies of fading appear to have been published.

Other Enhancement Processes in Quartz

The pink color of crystalline rose-quartz can sometimes be intensified by irradiation (A. J. Cohen, unpublished observation) and material that has been bleached by heating can also sometimes have its color restored. This color change is quite different from that in massive rose-quartz, where a smoky to black color develops on irradiation[23]. The color of rose quartz has been attributed to a wide variety of impurities and/or color centers. It is likely that it resides in charge transfer between the Ti^{4+} state of titanium and the Fe^{2+} of iron, induced by irradiation[24], just as in *Figure 3.5* and equation 3.15. Work is still actively underway on this problem. The suggestion has also been made[25] that a color center involving a phosphate impurity produces a second type of light-fading rose quartz. Indeed, Bauer[2] reports the rapid fading of rose quartz in light, while Frondel[1] states that 'The change in color, if any, is small'. **C**

Heating rose quartz produces a lightening of the color, which fades completely at 450 °C[1]. At least some such material can be restored by irradiation, with heat being used to remove any smoky quartz that may also form. **G**

Blue quartz derives its color from colloidal-size crystals of rutile, TiO_2[26], of other minerals[27], or merely from fine cracks[28]; all of these scatter light by Rayleigh scattering and thus produce the Tyndall blue color[26]. Heating to anywhere from 300 to 1000 °C can shift the color of some specimens toward violet, presumably by a small increase in the size of the scattering particles, can remove the color from others, presumably by a larger increase in the size, or can produce no change[1]. One might also expect that the asterism seen in some rose quartz could be modified by a heat treatment. **H**

Heating quartz by itself can produce cracking, particularly if done too rapidly when passing through the alpha-quartz to beta-quartz transition at 573 °C[1]. Some forms of quartz, particularly amethyst, acquire a milky 'schiller' when heated, for example, for 24 h at 550 °C[6], as shown in *Plate XVII,* and may then be used as moonstone imitations. **J**

Heating quartz and dropping it into a cold liquid produces fractures. This gives 'iris quartz' or 'rainbow quartz', showing interference colors within the cracks. Imitations of ruby, emerald, and so on can be made by filling these fractures with a colored dye solution, and are sometimes called 'crackled' or 'crackelée' quartz. This may be one of the oldest gemstone-enhancement processes known. It is mentioned in the 1500 year-old papyrus *P. Holm*, as discussed in Chapter 2, pp. 9–12, and was also described in detail by Boyle in 1672, as quoted on p. 18. Such dyed quartz is still seen periodically[28, 29] and is illustrated in *Plate XXIII.* **L**

Heating quartz that contains yellow-to-brown iron compounds can convert these into a red–brown to dark red color. This process is frequently used on tiger-eye to extend the range of available colors. **K**

The color of brown tiger-eye can also be lightened by bleaching with chlorine-containing bleaches or by removing the iron present in the tubes (from the decomposition of the original crocidolite asbestos) with a saturated oxalic-acid solution. Hydrochloric acid totally removes the filling of the tubes, converting tiger-eye into grayish cat's-eye[2]. The material that fills the tubes can also be dyed with aniline dyes where fading is possible, or with the stable inorganic impregnations described under chalcedony. **L**

A faceted, quartz doublet has been seen (S. Frazier, unpublished observation) that consists of a quartz front and back with a dendrite pattern drawn on the top surface of the back. Quartz triplets have been made to imitate emerald, ruby, and so on. **M**

Interference filters in the form of a surface coating[29] have been applied to faceted, colorless quartz to produce a deep yellow, citrine color or a green color somewhat resembling the daylight appearance of alexandrite. The coating may only be applied to the lower central region of the pavilion facets, but the color fills the stone by reflection; the color localization becomes evident on viewing such a stone from the side. **N**

Reflective foils are frequently used on rose or blue quartz to intensify the asterism derived from rutile needles[11, 30]. Other than this, the once frequently used colorless and colored foils are only rarely seen today. **O**

The identification of the variously enhanced quartzes of this section is only rarely required. Crackled quartz readily reveals its color localization on a careful microscopic examination, with immersion in a fluid being very helpful. The irradiated smoky quartz is indistinguishable from the natural smoky quartz. It is at times possible to identify the heat-treated citrine from material which has not been heat treated, since the polysynthetic Brazil-twinning almost always present in the original amethyst should be detectable; other tests may also apply[31], but it must be borne in mind that the distinction is of uncertain value, since amethyst has been converted into citrine or prasiolite by a natural heating at some localities[22], as described above under processes **E** and **F**. #

References

1. C. Frondel, Silica minerals, *The System of Mineralogy,* Vol 3, 7th Edn, Wiley, New York (1962)
2. M. Bauer, *Precious Stones,* Vol 2, pp. 471–495, 502–503, Dover Publications, New York (1968)
3. K. Nassau and B. F. Prescott, A reinterpretation of smoky quartz, *Phys. Status Solidi,* **A29,** 659–663 (1975)
4. K. Nassau and B. E. Prescott, Smoky, blue, greenish-yellow and other irradiation-related colors in quartz, *Min. Mag.,* **41,** 301 (1977)
5. A. J. Cohen and L. N. Makar, Models for color centers in smoky quartz, *Phys. Status Solidi,* **A73,** 593 (1982)
6. D. Chakraborty and G. Lehmann, Infrared studies of X-ray irradiated and heat treated synthetic quartz single crystals, *Neues Jb. Mineral., Mh.,* **1977,** 289 (1977)
7. K. Nassau, *Gems Made by Man,* Chilton, Radnor, PA (1980)
8. K. Nassau and B. E. Prescott, Growth-induced radiation-developed pleochroic anisotropy in smoky quartz, *Amer. Min.,* **63,** 230 (1978)
9. B. Sawyer, *Color Sorting of Irradiated Quartz Material,* US Patent 3 837 826, Sep 24 (1974)
10. J. Lietz and M. R. Hänisch, Über die Bildung von Farbenzentren im Quarz durch Elektrolyse, *Naturwissenschaften,* **46,** 67 (1959)
11. K. Simon, Beiträge zur Kenntnis der Mineralfarben, *Neues Jb. Miner. Geol. Paläont.,* **26,** 249 (1908)
12. K. Nassau, The effect of gamma rays on tourmaline, greenish-yellow quartz, pearls, kunzite, and jade, *Lap. J.,* **28,** 1064 (1974)
13. V. S. Balitsky, Synthetic amethyst: Its history, methods of growing, morphology, and peculiar features, *Z. dt. Gemmol. Ges.,* **29,** 5 (1980)
14. J. Sinkankas, Color changes in gemstones, *Lap. J.,* **17,** 616 (1963)
15. J. Lietz and W. Münchberg, Über die Färbung des Amethyst, *Neues Jb. Mineral., Mh.,* **2,** 25 (1957)
16. H. Bank, Citrin, *Z. dt. Gemmol. Ges.,* **25,** 189 (1976)
17. K. Nassau, Natural, treated, and synthetic amethyst–citrine quartz, *Lap. J.,* **35,** 52 (1981)
18. K. Nassau, Artificially induced color in amethyst citrine quartz, *Gems Gemol.,* **17,** 37 (1981)

19. J. I. Koivula, Citrine–amethyst quartz – a gemologically new material, *Gems Gemol.*, **16,** 290 (1980)
20. F. H. Pough, Greened amethyst, *Gemmologist,* **26,** 110 (1957)
21. J. Sinkankas, 'Green' amethyst from Four Peaks, Arizona, *Gems Gemol.*, **9,** 88 (1957)
22. T. R. Paradise, The natural formation and occurrence of green quartz, *Gems Gemol.*, **18,** 39 (1982)
23. E. G. Holder, The cause of color in rose quartz, *Amer. Min.*, **9,** 75 (1924)
24. G. Smith, E. R. Vance, Z. Hasan, A. Edgar, and W. A. Runciman, A charge transfer mechanism for rose quartz, *Phys. Status Solidi,* **A46,** K135 (1978)
25. D. Maschmeyer and G. Lehmann, Ein Strahlungdefect als Ursache der Färbung bestimmter 'Rosenquarze', *Z. dt. Gemmol. Ges.*, **31,** 117 (1982)
26. N. Jayaraman, The cause of the colour of the blue quartzes of the charnockites of South India and of the champion gneiss and other related rocks of Mysore, *Proc. Indian Acad. Sci.*, **A9,** 265 (1939)
27. R. T. Liddicoat, Jr, Dyed quartz, *Gems Gemol.*, **11,** 339 (1965)
28. R. V. Dietrich, Quartz – two new blues, *Min. Record,* **2,** 79 (1971)
29. H. S. Jones, *Enhanced Jewel Stones and Method of Forming Same,* US Patent 3 490 250, Jan 20 (1970)
30. C. F. Fryer (Ed.), Dyed 'crackled' quartz *and* star quartz, *Gems Gemol.*, **17,** 229 (1981)
31. G. Lehmann, Über die Färbungsursachen natürlicher Citrine, *Z. dt. Gemmol. Ges.*, **26,** 53 (1977)

Quartzite, *see* quartz

Rainbow Quartz, *see* quartz

Rhinoceros Horn, *see* ivory

Rhodochrosite, *see* calcite

Rhodolite Garnet

This material is reported by G. V. Rogers (unpublished observation) to change from its purplish color to a hessonite-type brownish color on heating. A brief check confirmed this change at about 600 °C in some rhodolites, but not in others; it could not be reversed by gamma irradiation.

Rock Crystal, *see* quartz

Rubellite, *see* tourmaline

Ruby, *see* corundum

Rutile

Pale yellow, synthetic rutile may be turned an attractive blue color by heating at over 1000 °C in a reducing atmosphere. Heating at 1000 °C in oxygen reverses this change[1].

Reference

1. K. Nassau, *Gems Made by Man,* p. 212, Chilton, Radnor, PA (1980)

Sapphire, *see* corundum

Sard, Sardonyx, *see* chalcedony

Satin Spar, *see* gypsum *and* marble

Scapolite

Colorless or yellow material, as well as scapolite cat's-eyes, were found to turn a purple (amethyst) color on irradiation by X-rays, electrons, or on exposure to radium[1, 2]. This strongly dichroic color faded rapidly on heating or on exposure to light, but was not stable even if the irradiated stones were kept in the dark. Naturally occurring scapolite of this color does not fade.

References

1. F. H. Pough and T. H. Rogers, Experiments in X-ray irradiation of gem stones, *Amer. Min.*, **32,** 31 (1947)
2. F. H. Pough, The coloration of gemstones by electron bombardment, *Z. dt. Ges. Edelsteink.*, **20,** 71 (1957)

Schorl, *see* tourmaline

Sea Palm Nut, *see* ivory

Selenite, *see* gypsum

Sepiolite, *see* marble

Serpentine, *see* marble

Shell, *see* pearl

Smithsonite, *see* calcite

Smoky Quartz, *see* quartz

Smoky Topaz, *see* quartz

Soapstone, *see* marble

Sodalite, *see* lapis lazuli

Sperm Whale Teeth, *see* ivory

Sphene

This material, also known as titanite, can be converted from a dark brown into a lighter orange or reddish brown by heating to red heat[1]. It does not seem to be known whether irradiation reverses this change.

Reference

1. J. Sinkankas, *Gem and Mineral Data Book,* p. 119, Winchester Press, New York (1972)

Spinel

Weigel[1] gives early references on the heat treatment of spinel and reports that purple and pink spinels change to yellow above 1000 °C, but may revert on cooling; blue spinel changes to green at 900 °C and to yellow above 1200 °C and this color change does not revert. Recently, red spinels were reported to have had their color improved by a heat treatment which removes a brownish component and leaves a pure red (R. Crowningshield, unpublished information).

Synthetic spinel[2] is usually grown by the Verneuil technique with an excess of Al_2O_3 to make growth easier. At the large excess composition, $MgO.5Al_2O_3$, the resulting clear product is unstable and fine crystallites of alumina precipitate on annealing. This forms a strong schiller, yielding a fine moonstone imitation[2, 3]. The refractive index of 1.73 is quite different from the 1.53 of moonstone, as is the specific gravity of 3.64 to 3.66 compared to 2.57.

Spinel triplets have been used to imitate emerald, ruby, and even tanzanite; fading can occur if the dyes used are not stable.

References

1. O. Weigel, Über die Farbenänderung von Korund und Spinel mit der Temperatur, *Neues Jb. Miner., Beil.*, **48,** 274 (1923)
2. K. Nassau, *Gems Made by Man,* pp. 246–249, Chilton, Radnor, PA (1980)
3. A. J. Breebart, Synthetic moonstone-colored spinel, *J. Gemm.*, **6,** 213 (1958)

Spodumene

Including: hiddenite and kunzite.

Summary of Treatments

Irradiation:
- **A** Pink to brown or green [u, f]
- **B** ? to yellow [s?]
- **C** Colorless to pink [s]

Heat:
- **D** Brown to green [u, f, r]
- **E** Green, purple, or pink to paler, colorless [s, r]

Spodumene, $LiAlSi_2O_6$, occurs in a variety of colors. When colored pink to purplish by manganese, usually with some iron also present, it is called kunzite. Chromium produces the emerald-green color of hiddenite. Other colors, such as yellow (iron?), yellow–green, and blue, have also been reported.

A Irradiation of pink or purple manganese-containing kunzite produces an intense, deep green color, as shown in *Plate II* which fades very rapidly (1 h or so) in light or on gentle heating. This is not a color center but involves the change from Mn^{3+} to Mn^{4+}, as described in Chapter 4, p. 58. This change was first reported in

1909 by Meyer; Przibram and Caffyn[1] give this and other early references. Almost any form of irradiation produces this change. Kunzite from Madagascar produces a brown color[1], believed to be a true color center, which fades extremely rapidly to the usual green color, which itself fades rapidly. **A**

Irradiation of spodumene of unknown type produces an intense orange to yellow to greenish yellow color, quite unlike any known in naturally occurring spodumene, and easily confused with citrine. This material, when offered in the trade as citrine, was found to be radioactive, indicating exposure in a nuclear reactor; the color was found to be stable to a short exposure to sunlight[2]. **B**

The color of heat-bleached kunzite (*see* process **E** below) can be restored by irradiation, which must be followed by heating or exposure to light to remove the green (and possibly brown) colors which form simultaneously, according to processes **A** and **B.** **C**

Heating of brown, irradiated spodumene to about 80 °C converts it into the green irradiated color, which itself is bleached to restore the original pink color at 200 °C[1]. Exposure to light also produces both of these changes. **D**

The pink color of kunzite fades on heating to about 500 °C, but can be restored by process **C** above, followed by process **D.** Here too, light has been noted to bleach the color of some kunzite, at least to some extent over extended periods of time[3]. Low temperatures, in the 100–250 °C range, are used to convert a bluish or purplish pink into a lighter, clearer pink color. **E**

The intense yellow and orange irradiated material of process **B** is unlike any spodumene found in nature; it does not have to be radioactive, since this depends on how the irradiation was performed. Green, irradiated kunzite colored by process **A** is sometimes seen[4]. Its color is sufficiently different from that of hiddenite, so that distinction should be no problem[4]. Merely leaving it for a few hours in a well-lit location produces evidence of the fading. #

References

1. K. Przibram and J. E. Caffyn, *Irradiation Colours and Luminescence,* pp. 246–250, Pergamon Press, London (1956)
2. G. R. Rossman and Y. Qiu, Radioactive irradiated spodumene, *Gems Gemol.,* **18,** 87 (1982)
3. J. Sinkankas, Color changes in gemstones, Part Four, *Lap. J.,* **17,** 616 (1963)
4. R. T. Liddicoat, Jr, Irradiated spodumene and morganite, *Gems Gemol.,* **12,** 315 (1968)

Star Corundum, Ruby, Sapphire, *see* corundum

Stars, *see individual gemstone variety involved*

Steatite, *see* marble

Succinite, *see* amber

Sugilite

It has been noted by J. Call (unpublished observation) that dark purple sugilite can be lightened by heating at about 450°C; the product may have a banded appearance.

Sunstone, *see* feldspar

Talc, *see* marble

Tanzanite, *see* zoisite

Tiger-Eye, Tiger's Eye, *see* quartz

Titanite, *see* sphene

Topaz

Summary of Treatments

Irradiation:
A Colorless or pink to yellow–brown (BSCC) [s, r]
B Colorless or pink to yellow–brown (BFCC) [u, f, r]
C Colorless to brown or greenish [s or u, f, r]
Heat:
D Brown or orange to pink, 'pinking' [s, r]
E Yellow, green, or brown to colorless or blue [s, r]
F Blue to colorless [s, r]

Topaz, $Al_2SiO_4(F,OH)_2$, occurs in a wide range of colors, including yellow, orange, brown, pink-to-violet, and blue. All of these are due to color centers, with the exception of the pink-to-violet and the pink component of some orange, which are caused by a chromium impurity (similar to the chromium-caused red in ruby).

It is one of the characteristics of certain yellow-to-brown colorations that merely changing the intensity appears to change the color. This is true of topaz, where intensification produces the sequence yellow to orange to sherry brown to a deep reddish brown. When brown is applied to topaz herein, any of these colors are meant to be implied. Similarly, BFCC is used for brown-fading color center and BSCC for brown-stable color center, to distinguish these two sets of colors.

Much natural brown topaz fades in the light[1–3]. The color may be lost in a few days in sunlight as, for example, in the yellow-to-brown material from Utah and from some Mexican locations; these contain the BFCC as the cause of color.

A,B Almost all topaz turns a brown color on irradiation as in *Figure 7.14(b)*. This was observed in the earliest experiments with radium salts and subsequently with all other energetic irradiations[4]. In a detailed study[5, 6] of 85 specimens, this brown coloration was found to form at two different rates; the rapidly forming BFCC reddish brown color also faded rapidly within a few days in light, while the slowly forming BFCC darker brown faded much more slowly, in a few weeks. In Petrov's specimens[1] the BFCC was reddish brown while the non-fading BSCC was a yellowish shade, which could have been merely an intensity difference.

Not all irradiation-produced colors fade; this apparently depends on the specific ingredients (of unknown types) present, some of which can stabilize the color centers. When natural brown topaz is bleached by heating and then has its color restored by irradiation, the originally fading material still fades; the originally stable material may give a darker color than the original because both BSCC and BFCC are formed, but after light fading the original color is returned. This provides a means of restoring the color if a natural brown topaz is overheated. **A,B**

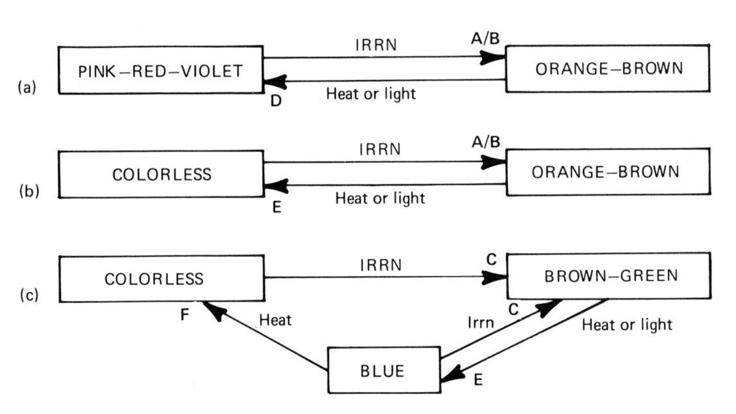

Figure 7.14 The heat and irradiation changes involving topaz: (a) chromium containing; (b) brown only; (c) brown and blue. The letters **A** to **E** refer to the processes of the topaz summary table

Colorless natural topaz usually forms only the BFCC, although material from some locations is also reported to form the BSCC. Two different shades of the fading brown BFCC are shown in *Plate XII.*

Pink topaz can be turned brown by this same process. A single report[7] mentions the production of a reddish pink color by the neutron irradiation of colorless topaz, but without any further details as to origin, stability, and so on.

Much topaz turns a brown-to-green color on irradiation. This material has the property that heating or light exposure now removes the yellow or brown component and leaves a blue color, as in *Figure 7.14(c)*. Pough[4] appears to have been the first to observe this process, with details later reported by the author and co-workers[5, 6] and others[8]. This blue color is stable to light and two shades of it are shown in *Plate XII.* **C**

A curious phenomenon here is that merely extending the irradiation time in a gamma cell does not seem to increase the depth of the blue color beyond a certain point. Special techniques need to be used to produce an intense, deep blue color, much deeper than that of natural blue topaz. It appears that a very intense irradiation is required, such as that available in an electron accelerator, operated at a high energy, or in a nuclear reactor. Both of these can induce radioactivity by activating certain impurities, as in some blue stones that were seen in the trade[9] and are described in Chapter 4. It is reported that this can be avoided by using only that colorless topaz which does not contain activating impurities. It is also sometimes mentioned that heating during irradiation assists (or prevents) the formation of the deep blue. So far there are no published details on these subtleties. Very large doses are reported to be used, as high as 40 000 Mrad for some types of topaz.

Brown topaz from certain localities where chromium is present passes through orange to a pink or violet color on being heated, as in *Figure 7.14(a)*. This 'pinking' **D** process produces a stable color which is not affected by further heating. It was at one time performed with the material buried in sand, etc.[2], to 500 °C (*see also Table 3.1*), but today is usually done by gently heating in a test-tube over an alcohol burner, for example, so that the process can be watched and stopped just at the right point. Irradiation restores the original color by processes **A** or **B**. **D**

Much topaz loses its brown color on heating at any temperature from 200 to over 400 °C for a few hours and turns colorless[1, 5, 6], as in *Figure 7.14(b)*. Processes **A** or **B** restore the color. Topaz irradiated by process **C** is similarly heated to reveal the underlying blue color, as in *Figure 7.14(c)*. This can be reversed by processes **A** or **B**. **E**

The blue color of topaz, whether natural or derived from irradiation, can be removed by heating to about 450 °C, as also shown in *Figure 7.14(c)*. This too can be reversed by processes **A** or **B**. A 'steely' blackish-blue color can be softened by a short heating to about 230 °C. **F**

The identification of the various shades of pink, brown, and blue presents some difficulties. The 'burned' pink is reported to have a much larger dichroism (dark cherry red and honey yellow) than the very rare, natural pink topaz[2]. Both the irradiated brown[5, 6, 10] and the blue[5, 6, 8] colors show identical absorption spectra to the natural stones, but there are other differences. The irradiated browns usually show the lower refractive index of colorless topaz, 1.61–1.62, as against the 1.63–1.64 of natural brown topaz and, similarly, a lower specific gravity, 3.52–3.53, as against 3.56–3.57[11]. With the blue material there is a difference in the thermoluminescence as reported by Petrov *et al.*[10]; as the temperature is raised, there is a large emission of light at about 360 °C in the irradiated blue, not present in the natural blue. This could be performed on a microscopic specimen scraped from the girdle of a stone, but would require extensive testing to ensure that it is a consistent and reproducible test which could not be negated by a selective heating of the stone. #

Finally, for the yellow-to-brown shades of topaz, there is the problem of distinguishing the stable BSCC from the fading BFCC. At present no gemological test is known, other than a direct fade test, for this identification.

References

1. I. Petrov, Farbe, Farbursache und Farbenveränderungen bei Topasen, *Z. dt. Gemmol. Ges.*, **27**, 3 (1978)
2. M. Bauer, *Precious Stones*, p. 332, Dover Publications, New York (1968)
3. J. Sinkankas, Color changes in gemstones, Part Four, *Lap. J.*, **17**, 616 (1963)
4. F. H. Pough, The coloration of gemstones by electron bombardment, *Z. dt. Ges. Edelsteink.*, **20**, 71 (1957)
5. K. Nassau and B. E. Prescott, Blue and brown topaz produced by gamma irradiation, *Amer. Min.*, **60**, 705 (1975)
6. K. Nassau, The effects of gamma rays on the color of beryl, smoky quartz, amethyst and topaz, *Lap. J.*, **28**, 20, correction p. 556 (1974)
7. K. S. Raju, Topaz – on neutron irradiation, *Int. J. appl. Radiat. Isotopes*, **32**, 929 (1981)
8. I. Petrov and W. Beredinski, Untersuchung künstlich farbveränderter blauer Topase, *Z. dt. Gemmol. Ges.*, **24**, 16 (1975)
9. R. Crowningshield, Irradiated topaz and radioactivity, *Gems Gemol.*, **17**, 215 (1981)
10. I. Petrov, W. Beredinski and H. Bank, Bestrahlte gelbe und rotbraune Topase und ihre Erkennung, *Z. dt. Gemmol. Ges.*, **26**, 148 (1977)
11. R. T. Liddicoat, Jr, Irradiated topaz, *Gems Gemol.*, **12**, 155 (1967)

Topaz, Smoky, *see* quartz

Tortoise Shell

This material is the carapace or upper shell of the marine Hawksbill turtle. Heat (boiling water) and pressure are used to flatten this thermoplastic material and freshly scraped edges can be joined in the same way; the color may be darkened by this process (Webster, pp. 598–602). Layers of this shell may be similarly laminated to form beads[1] and other solid objects, where swirls and contact planes can be seen. The same is true of reconstituted tortoise shell, made from small pieces in a similar way as is reconstructed amber; coloring may be added in the process[1]. Tortoise shell may be cemented to sheet plastic for bulk and strength.

Reference

1. R. Crowningshield, Laminated tortoise shell, *Gems Gemol.*, **11**, 366 (1965–1966)

Tourmaline Group

Including: achroite, buergerite, dravite, elbaite, indicolite, indigolite, liddicoatite, rubellite, schorl, uvite, and verdelite.

Summary of Treatments

Irradiation:
A Pale colors to pink or red, blue or dark green to purple, yellow to peach; all [s or f, r?]
B Pale colors to yellow, blue–green to yellow–green, pink to orange; all [s or f?, r?]
Heat:
C Red to lighter to colorless, brownish red to pink, purple to blue or dark green [s, r?]
D Dark blue and green to lighter blue–green or yellow–green [s]
Other:
E Foil, acid treatment, impregnations [s or u]

The name tourmaline includes buergerite, dravite, liddicoatite, schorl, and uvite. The general formula, $(Na,Ca)(Li,Mg,Al)_3Al_6B_3Si_6O_{27}(OH,F)_4$ indicates the complexity of the family. In the words of John Ruskin[1]: 'the chemistry of it is more like a medieval doctor's prescription than the making of a respectable mineral'. Transparent gemstones of green, sometimes called verdelite, red, sometimes called rubellite, blue, sometimes called indicolite or indigolite, black, usually but not necessarily schorl, and yellow-to-orange-to-brown colors may occur in several of these family members and would normally require chemical analysis for a correct assignment.

The causes of color in the tourmalines are equally complex, derived from the fact that not all stones of any given color necessarily have the same color origin. In general it can be said that iron causes the greens and blues (with charge transfer involved at times) as well as yellow-to-browns and part of the pinks and reds, but manganese is also involved in the pinks and reds[2, 3]. As might be expected from all of this complexity, the results of enhancement treatments can be varied and unpredictable; they should be consistent, however, for any given locality.

Irradiation has two major effects: it can develop or intensify either a red or a yellow color[2, 4–6]. Out of a large group of specimens tested, 193 developed or intensified red, 70 developed or intensified yellow, and 274 did not change[4]. These changes are illustrated in *Plates IV* to *VI*. **A,B**

Very pale pink, green or blue tourmalines frequently become a deeper pink to a red on irradiation; blue and green stones may turn a deep purple; yellow stones may turn an orange or peach color; and medium green stones may turn a gray (if the green and red absorptions just balance) or even convert to a green–red bicolor. The effect of this treatment on five bicolor tourmalines has been reported[4]. Most of these red components are stable on heating, even to over 400 °C, while some lose their red component by 260 °C[2, 4]; only these latter specimens lost some of their color after several weeks in sunlight[4]. Although cracking during electron bombardment to produce red was said not to be due to heating[5], other techniques do not produce cracking; the intensity of the heating effect of high-energy electrons in such a poor thermally conducting material was undoubtedly underestimated. All these changes can probably be reversed by heat and are suspected to involve valence state changes of manganese and/or iron impurities; color centers are probably not involved. **A**

Irradiation may develop or intensify a yellow color, an original pink or red color may disappear or produce orange, while a blue or a green may turn bluish green or even yellow[4, 5]. There is relatively good stability to heating and no light bleaching was noted[4]; the valence states of manganese and/or iron may be involved. **B**

Heating removes the pink or red component from tourmalines, producing colorless stones from pinks and reds, blues from some browns and purples, and yellows from orange stones. Some irradiated pink or red color disappears by 260 °C (this type may also fade in light), some more by 400 °C, and all decolorizes by 700 °C (Webster, p. 150). One could expect this change to be reversed by irradiation, but such experiments do not seem to have been reported. Brownish stones have also been reported to turn into pink stones, apparently losing their off-color component at relatively low temperatures. **C**

Long ago, it was observed[7] that blue–green tourmalines can be lightened by heating to 650 °C; *see also Table 3.1,* p. 26. Temperatures from 500 °C to almost 750 °C have been reported for lightening green or blue–green stones, but great care must be taken, because essential water is lost with destruction of the tourmaline at about 725 °C[7] (this temperature can be expected to vary somewhat with the composition). Depending on the oxidizing or reducing nature of the atmosphere used, changes from green to pink or from pink to colorless or slightly bluish, all at 500 °C, have also been reported[8]. In the absence of detailed studies, it is difficult to draw general conclusions from these fragmentary reports, except to be reasonably **D** sure that some, but not all of the blue and blue–green stones can be made greener and/or lighter, at least to some extent, and that the product is almost certainly stable to light. **D**

Foils were used in the past to brighten dark tourmalines. Colorless wax is used to fill small cracks and improve the appearance in the polish of tumbled and carved tourmalines. Tourmaline cat's-eyes are sometimes acid treated to lighten the color by removing dark filling from the tubes that cause the eye effect; wax (or plastic?) may then be used to fill the tubes to prevent the entry of dirt (R. Crowningshield, unpublished observation). **E**

No identifying characteristics are known for the irradiated or heat-treated tourmalines. #

References

1. J. Ruskin, *The Ethics of the Dust; Ten Lectures to Little Housewives on the Elements of Crystallization,* London (1890)
2. K. Nassau, Gamma ray irradiation induced changes in the color of tourmalines, *Amer. Min.,* **60,** 710 (1975)
3. E. Althaus, Wassermelonen und Mohrenköpfe, *Lapis,* **4,** 8, 48 (1979)
4. K. Nassau, The effect of gamma rays on tourmaline, greenish-yellow quartz, pearls, kunzite, and jade, *Lap. J.,* **28,** 1064 (1974)
5. F. H. Pough and T. H. Rogers, Experiments in X-ray irradiation of gem stones, *Amer. Min.,* **32,** 31 (1947)
6. F. H. Pough, The coloration of gemstones by electron bombardment, *Z. dt. Ges. Edelsteink.,* **20,** 71 (1957)
7. Anon, Improving the colour of tourmalines, *Gemmologist,* **1,** 213 (1932)
8. P. L. C. Grubb and T. H. Donnelly, Colour changes in elbaite tourmaline from Ravensthorpe, Western Australia, *Austral. Gemm.,* **10,** 15 (1969)

Travertine, *see* marble

Turquoise

Summary of Treatments

A Colorless impregnation, 'stabilization' [s or u]
B Colored impregnation [s or u, f]
C Dyeing [s or u, f]
D Coating, colorless or colored [s or u, f]
E Color 'restoration' [u]

Turquoise is an aggregate of tiny crystals of copper aluminum phosphate hydrate, $CuAl_6(PO_4)_4(OH)_8.5H_2O$. The blue color derives from the idiochromatic copper content. Specimens range from compact to porous; with high porosity there may be so much scattering of white light from the pore structure that the blue color cannot be seen. If the pores are filled with a substance having a higher refractive index than air (e.g. water, oil, wax, plastic), then the inherent blue color of the turquoise usually becomes visible. The fading of turquoise frequently referred to in older works[1] is mostly the drying out of moisture from these pores. Attack by acids and other chemicals present in some beverages, in skin oils, in the perspiration of some individuals, and in many cosmetic products can change the ligand field about the copper in the surface of the turquoise and produce a change in color from blue to green. The susceptibility to this change obviously depends on the degree of porosity present.

A The pores of turquoise can be filled by oil, wax, plastic, or a number of other substances. This widely used process is frequently referred to as 'stabilization' since it prevents or reduces the color changes induced by the agents described above. Other terms used include treatment, color-stabilization, permanizing, conversion, and so on. Fat and mineral oil were much used at one time, but these are not very stable in themselves and some may also produce a color change. Wax impregnation, in particular with paraffin wax, is frequently seen[2]. The turquoise is first gently dried and then soaked in warm, melted wax, possibly over a double boiler[3], for several days; the use of a vacuum or of pressure is sometimes mentioned for such impregnations, but is probably not necessary[3, 4]. Plastics, of the epoxy type or others, have been frequently used[5, 6] and have much more permanence than does wax.

Inorganic substances have also been used as impregnants, including soaking in a waterglass (sodium silicate) solution followed by concentrated hydrochloric acid to form a silica gel[7, 8] within the pores, and the application of a colloidal dispersion of silica in water[8]. Such silica impregnations lighten the color of the turquoise in the absence of an added colorant.

B Impregnation with colored agents is sometimes employed; this process is distinct from dyeing, where the solvent used evaporates. Most frequently a blue wax[9] or a blue plastic is applied, in the same way as in process **A.**

C Dyeing of turquoise is only rarely seen, sometimes being combined with one of the other processes. An old English patent cited by Pogue[10] played it safe by combining two colorants: a copper or iron tartrate deposition followed by an aniline dye! Prussian blue, applied as described for chalcedony in Chapter 5, p. 70, can also be employed[10]. The deep blue cuprammonium solution, made from saturated copper sulfate with an excess of ammonium hydroxide, has been used[8] to dye the colorless silica impregnation of process **A**, as has also been done with a copper–amine complex, made from saturated copper sulfate mixed with an excess of monomethylamine[7]. Dye has been applied selectively to cover white areas in turquoise[11].

D Coatings, either colorless or colored, are reported surprisingly often. Colorless coatings are sometimes applied to protect an underlying dye[12], impregnation, or paint layer. Sometimes the surface is etched with acid, a blue epoxy resin is applied, and the excess removed by polishing[13]. In a three-step combination coating[14], the surface of turquoise beads was painted blue, the appearance of matrix was added with black paint, and the whole covered with clear lacquer; one wonders why turquoise was used for the beads at all! Finally, in the blue wax impregnation of process **B**, a statue had matrix veins painted on and much blue wax was left as a surface coating to hide the greenish color of the turquoise[9].

E Processes for 'restoring' turquoise altered by grease or chemicals have sometimes been reported. Some sources mention ammonia[1] or hydrogen peroxide[9] for restoring the blue color, but this treatment is only temporary, if it works at all. The use of solvents to remove oil or wax has also been investigated, employing gasoline[15] or ether (very dangerous and a rather poor solvent choice) used in a Soxhlet continuous-extraction device to remove face and suntan preparations[16]. It is doubtful if any of these restoration processes really work. **E**

The prime detection device for treatments, once synthetics and imitations have been eliminated, is a careful microscopic examination. Most surface coatings become evident, as does most dyeing, from the localization of color in cracks. Ammonia applied on a white swab often removes traces of dye, at least in the absence of a protective coating. The hot point applied with care usually reveals a wax coating by local melting and may reveal a plastic coating by melting and/or a characteristic odor[6]; some plastics do not give either of these reactions[5]. An infrared-absorption test has been suggested to be useful at times[17]. #

The colorless impregnation 'stabilization process' is widely accepted in the trade and not usually disclosed, as discussed in Chapter 6.

References

1. M. Bauer, *Precious Stones,* p. 391, Dover Publications, New York (1968)
2. L. B. Benson, Old turquoise oiling method still used, *Gems Gemol.,* **6,** 221 (1949)
3. Anon, Stabilizing, *Int. Turquoise Ann.,* **2,** 83 (1977)
4. Anon, It's not that simple, *Int. Turquoise Ann.,* **2,** 86, 91 (1977)
5. R. Crowningshield, Plastic-treated turquoise, *Gems Gemol.,* **13,** 117 (1969–1970)
6. R. T. Liddicoat, Jr, Unusual turquoise treatment, *Gems Gemol.,* **15,** 139 (1976)
7. F. B. Wade and W. C. Geisler, Staining turquoise, *Gemmologist,* **18,** 187 (1949)
8. H. Lee and R. Webster, Imitation and treated turquoise, *J. Gemm.,* **7,** 249 (1960)
9. R. T. Liddicoat, Jr, Paraffin-treated turquoise, *Gems Gemol.,* **12,** 152 (1967)
10. J. E. Pogue, *The Turquoise,* Vol 12, Memoir 3, p. 133, National Academy of Sciences, Washington, DC (1915)
11. R. T. Liddicoat, Jr, New form of turquoise treatment, *Gems Gemol.,* **14,** 25 (1972)
12. R. Crowningshield, Dyed, plastic-treated turquoise, *Gems Gemol.,* **13,** 118 (1969–1970)
13. L. B. Benson, Turquoise, *Gems Gemol.,* **10,** 51 (1960)
14. R. Crowningshield, Doctored turquoise, *Gems Gemol.,* **11,** 359 (1965–1966)
15. Anon, Synthesen, Imitationen, Färbungen, *Z. dt. Ges. Edelsteink.,* **56,** 26 (1966)
16. V. Theisen, Regenerien von Türkisen mit der Soxhlet-Apparatur, *Z. dt. Ges. Edelsteink.,* **53,** 20 (1965)
17. T. Lind and K. Schmetzer, Zur Bestimmung natürlicher, behandelter, and synthetischer Türkise sowie von Türkisimitationen, *Z. dt. Gemmol. Ges.,* **32,** 69 (1983)

Uvite, *see* tourmaline

Vegetable Ivory, *see* ivory

Verd-antique, *see* marble

Verdelite, *see* tourmaline

Verdite, *see* marble

Vulpinite, *see* celestite

Walrus Tusk, *see* ivory

Wollastonite

White, compact wollastonite, $CaSiO_3$, has been dyed a green color to imitate jade[1].

Reference

1. R. T. Liddicoat, Jr, Developments and highlights, *Gems Gemol.,* **15,** 295 (1977)

Wood, Petrified, *see* chalcedony

Zircon

Including: hyacinth, jacinth, jargon, jargoon, etc.

Zircon, $ZrSiO_4$, comes in a variety of colors, including greenish blue, colorless-to-pale yellow (sometimes called jargon or jargoon), and the most commonly found orange or reddish brown (hereafter called merely brown; sometimes called hyacinth or jacinth). At one time colorless zircon was much used as a diamond imitation, but has been replaced by the much superior synthetics.

As mentioned in Chapter 4, pp. 52–53, zircons frequently contain significant quantities of radioactive thorium and related elements[1]; over geological periods this internal irradiation can produce sufficient atomic displacements within the zircon to change its properties. The disordered state produced by irradiation is called the metamict state. The specific gravity of about 4.70 in 'normal' or 'high' zircon can thus be reduced to about 3.95 in 'metamict' or 'low' zircon, with 'intermediate' zircon in between; the analogous refractive index values are 1.92, 1.98 for high and 1.81, 1.81 for the low zircon, and the lattice parameters 6.60, 5.98 Å for high and 6.71, 6.09 Å for low zircon[2]. Low zircons are mostly green and cloudy, the latter derived from the separation of the $ZrSiO_4$ into the separate components ZrO_2 and SiO_2. Intermediate zircons on the high side can be heat treated, typically in the 1000–1400 °C range, to restore the high state. The lowest zircons are usually those containing the highest quantity of radioactive ingredients and cannot be converted, even at 1450 °C, in a reasonable time. References to early studies and a good summary have been given by Holland and Gottfried[2].

Zircon gemstones are basically all in the high state. There are some changes in color in the brownish 'hyacinth-type' zircons on being heated and exposed to light or stored in the dark, summarized in *Figure 7.15(a),* based on the work of Wild[3]; these changes are minor and not important, since these unattractive colors are rarely seen in the trade.

The drastic changes in color of zircons aimed at producing attractive, colorless, blue, and yellow-to-red colors are summarized in *Figure 7.15(b)*. Descriptions of the practices in Thailand have been given by Buckingham[4] and by Eppler[5], who also give many early references. The brown zircons are first heated while surrounded by charcoal to about 1000 °C for a few hours. This reducing treatment converts some of the stones to blue, some to colorless, and leaves some off-color. After a possible repeat of this treatment, the off-color stones are next heated to about 900 °C with access to air. This oxidizing treatment converts some to colorless and some to yellow, orange, or red. Some stones may remain off-color and may be recycled through either process. There is the possibility that a very slow cooling after the oxidation heating is beneficial. The addition of chemicals has sometimes been mentioned (for instance, cobalt nitrate or potassium ferrocyanide[6]), but this is almost certainly never used.

Just as natural zircons may change color in the light and revert in the dark[7], so some of the heated colors have been reported to alter with time or on exposure to light. Treated stones are reported to be exposed for several days to the sun or stored for up to one year in the dark so that unstable stones can be removed; retreatment is usually tried, although this may not help. The radioactivity of

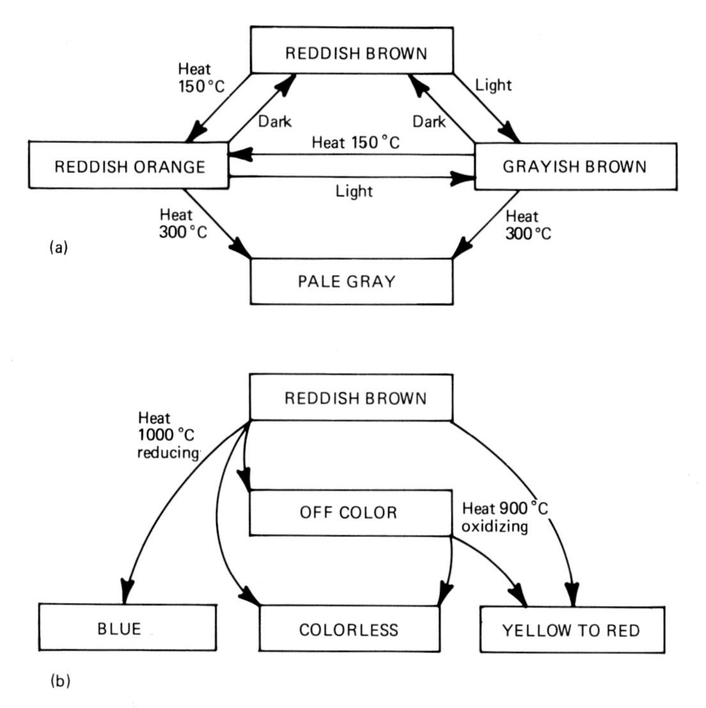

Figure 7.15 (a) The low temperature and (b) high temperature color changes that occur in zircon; temperatures in °C

gem-grade high zircon is usually extremely low, and alteration due to internal activity probably requires many centuries. Irradiation usually restores the original reddish or greenish brown color, should this be desired.

The identification of colorless, blue, and red zircon is not normally attempted, # it being automatically assumed that such material has been treated. Some stones of these colors do occur in nature, but only rarely, and such material is not seen in the gemstone trade (R. Crowningshield, unpublished observation).

References

1. W. A. Deer, R. A. Howie, and J. Zussman, *Rock Forming Minerals,* Vol 1, pp. 59–68, Wiley, New York (1962)
2. H. D. Holland and D. Gottfried, The effect of nuclear radiation on the structure of zircon, *Acta Cryst.,* **8,** 291 (1955)
3. G. O. Wild, Colour changes in zircon, *Gemmologist,* **7,** 98 (1938)
4. W. C. Buckingham, The mining and heat treatment of zircons, *J. Gemm.,* **2,** 177 (1950)
5. W. F. Eppler, Das Geheimnis des Zircons, *dt. Goldschmiede Zeitung,* **51,** 531 (1936)
6. E. S. Dana and W. E. Ford, *A Textbook of Mineralogy,* 4th Edn, p. 610, Wiley, New York (1966)
7. M. Bauer, *Precious Stones,* Vol 2, p. 342, Dover Publications, New York (1968)

Zoisite

The mineral zoisite, $Ca_2Al_3Si_3O_{12}OH$, also occurs in a gem form called tanzanite, discovered only in 1967[1]. Although a few, deep violet–blue dichroic crystals have been found, the majority of this material occurs with a brownish color having a strong violet–red/deep blue/yellow–green trichroism. Heating for about 2 h to 370 °C converts the yellow–green component into a deep blue, leaving only a violet–red/deep blue dichroism and the desired color closely resembles sapphire, as shown in *Plate XVIII*. Some material is said to require a temperature over 600 °C. The color is stable to 900 °C, but above this temperature water is lost and a dirty yellow results. The causes of the color and color change have not yet been established with certainty; Gübelin and Weibel have summarized this and the gemology[2].

References

1. C. S. Hurlbut, Jr, Gem Zoisite from Tanzania, *Amer. Min.*, **54**, 702 (1969)
2. E. Gübelin and M. Weibel, Neue Untersuchungen am blauen Zoisit (Tanzanit), *Z. dt. Gemmol. Ges.*, **25**, 23 (1976)

More on Heating

Some details of heating for the enhancement of gemstones were given in Chapter 3; it is assumed that the reader has covered that material. The simplest way of heating at a low temperature is in a glass (pyrex) test-tube held over an alcohol, gas, or propane flame, as there described. Next there is the home baking-oven, usually with a limit of about 300 °C (575 °F). A charcoal brazier can reach a much higher temperature, particularly if bellows are used to supply extra air, but the lack of temperature control prevents this from being used for any but the most primitive of heat treatments. In this appendix are considered somewhat more sophisticated furnaces, heated either electrically or by a gaseous fuel, where good temperature-control is possible and consistent heat-treating results can be expected, given suitable raw material.

A WARNING: *Many of these furnaces present hazards, not only from the high temperature, but also from the high voltage and explosive or toxic gases which may be used or produced. Furnace construction and operation should be attempted only by a person with an adequate background to anticipate the hazards involved. The instruction manuals for commercial furnaces should be read for their cautionary warnings.*

Furnace Construction

Information on furnace construction is widely scattered. There are some specialized books, such as *Industrial Electric Furnaces and Appliances*[1], *The Science of Flames and Furnaces*[2], and the article by Start and Thring[3], as well as books on temperature measurement such as that by Kinzie[4]. Important information is found in texts on ceramics such as the one by Kingery, Bowen, and Uhlmann[5] as well as in specialized books such as *High-Temperature Materials and Technology*[6]. Useful sections on furnace construction and temperature control are included in the German text on crystal growth by Smakula[7]. Extremely useful are some of the data sheets and catalogs of manufacturers on ceramics and heating elements, such as those of the Norton[8] and Kanthal[9, 10] Companies, and on temperature measurement and control, such as that of the Omega Company[11] (*see also* Appendix D).

Except for the very lowest- and highest-temperature furnaces, the useful hot region inside a furnace is usually bounded by some type of ceramic partition. In

Table A.1 are listed furnace-construction materials grouped according to type, and by maximum working temperature within each type. The most commonly used materials are the ceramic oxides, which are available in a wide range of shapes including plates, disks, tubes, thimbles (tubes closed at one end), cores (tubes or thimbles with a spiral groove on the outside), and so on. It must be recognized that

Table A.1 Crucible and furnace materials

Material	*Max. useful temp.* (°C)	*Comments*[a]
Glasses		
Pyrex glass	500	ONR
Vycor glass	1100	ONR
Silica glass (fused quartz)	1200	ON(R)
Oxides		
Porcelain	1100–1300	ONR
Steatite, talc	1250	ONR
Firebrick, fireclay	1200–1500	ONR
Firebrick, fireclay, high alumina	1600	ON(R)
Mullite ($3Al_2O_3.2SiO_2$)	1700	ON(R)
Sillimanite ($Al_2O_3.SiO_2$)	1700	ON(R)
Zircon ($ZrO_2.SiO_2$)	1750	ON(R)
Spinel ($MgO.Al_2O_3$)	1800–1900	ONR, shock
Alumina (Al_2O_3)	1850–1950	ONR
Magnesia (MgO)	2300	O, shock
Zirconia, stabilized (ZrO_2+)	2300	ON(R), shock
Thoria, stabilized (ThO_2+)	2700	ON, shock
Metals		
Iron, nickel	1100	(O)NR
Platinum	1500	ONR
Rhodium	1800	ONR
Tantalum	2000	NR
Iridium	2100	ONR
Molybdenum	2100	NR
Tungsten	3000	NR
Other		
Silicon carbide (SiC)	1500	ON
Silicon nitride (SiN)	2900	ON
Carbon, graphite	3000	NR

[a] Usable in oxidizing, O, neutral, N, or reducing, R, atmospheres; 'shock' indicates particularly poor resistance to thermal shock, that is to sudden temperature changes.

there is considerable uncertainty in the temperatures listed in *Table A.1*. Thus, the 1950 °C listed as the maximum useful temperature for alumina, the most commonly used furnace-construction material, represents the very highest temperature that can be used for limited periods with the very purest of alumina ceramic that is over 99.8 percent pure. It is necessary to reduce this temperature by 100 °C or more for extended periods and even further for lower purity material.

The maximum temperature that can be used also depends on the stress applied to a material. Thus, if there is no particular loading on a pure alumina member, the 1950 °C of *Table A.1* is just feasible. If the alumina is under a significant load, then another 100 °C or more must be subtracted from the maximum temperature. If deformation at temperature can be tolerated, however, then the full limit may be feasible, even in the presence of some stress.

The comments given in the last column of *Table A.1* also need to be interpreted with care. Most oxides that operate to their temperature limit in oxygen or oxidizing atmospheres, such as air, can also be used in neutral atmospheres, such as vacuum, nitrogen, argon, etc., and tolerate reducing atmospheres, such as hydrogen or carbon monoxide, quite well. If some deterioration is possible at the highest temperatures, then the symbol is given in parentheses. Thus, iron may be used up to 1100 °C in an inert or reducing atmosphere, but begins to react with oxygen above about 750 °C.

Second only in importance is the furnace insulation used to surround the hot region. This not only reduces the heat loss and the fuel or energy cost, but also permits operation close to the limit of the heat source. The best insulators are light, fluffy powders, loose fibers, or felted fibers that entrap the air, which actually does the insulating. These have the lowest values of the thermal conductivity, as shown in *Table A.2*. The value for solid magnesia is also included in this listing to illustrate

Table A.2 Furnace-insulation materials

Material	*Max. usable temp.* (°C)	*Thermal conductivity*[a]
Glass fiber	600	0.03
Silica fiber	1000	0.1
Firebrick, insulating	1200–1500	0.3
Fiberfrax	1650	0.07
Alumina, bubble	1800	0.6
Magnesia, powder	2200	0.3
Magnesia, solid	2300	1.7
Carbon or graphite, powder[b]	3000	0.05
Radiation shields, molybdenum[b]	2100	0.4

[a] Units of BTU/(h ft^2) (°F/ft); the lower the value, the better is the thermal insulating effect for equal thickness.
[b] Requires inert or reducing atmosphere.

the change which is produced by powdering. Carbon powder and molybdenum or other metal radiation-shields are usually employed only at the very highest temperatures and require operation in an inert or reducing atmosphere. The use of such materials does not, however, limit the atmosphere surrounding the sample, since an extra protective tube or container can separate the reducing atmosphere required for the heating element from the working space where the sample could be in any desired atmosphere, as described below.

Increasing the amount of thermal insulation also permits the outer shell of the furnace to remain cooler. In some very high temperature furnaces it may not be practical to use a large amount of insulation, and the outer furnace shell may then have water circulating through it to keep it cool.

The materials used for furnace walls or insulation can also be employed for the crucibles that hold the specimen. Here it is important to have compatibility between the crucible, the specimen, any other chemicals present, and the atmosphere. At times other factors have importance, as exemplified by the apparently arbitrary statement sometimes heard that the heat treatment of blue sapphire works in an alumina crucible, but not in a magnesia crucible. The reason for this may derive from the high volatility of magnesia at high temperatures and some specific chemical interaction between the magnesia vapor and the corundum.

Within the hot region of the furnace three mechanisms can occur for heat transfer: conduction, convection, and radiation. At the lower temperatures, conduction is active, but the main heat-transfer mechanism operates by convection of the hot gas within the furnace. This is a rather slow process and considerable time (several to tens of minutes) is required for a specimen to reach the full furnace temperature. Radiation also becomes active at the higher temperatures; this is a very rapid heat-transfer mechanism and a specimen placed into a furnace at, say, 1700 °C equilibrates to this temperature in just a few minutes. More commonly, specimens need to be placed into the cold furnace for a slow warming-up, to avoid the cracking described in Chapter 3.

Resistance-heated Furnaces

The most commonly used furnaces employ a conducting element that has a medium-to-high resistance to generate heat from the flow of an electric current. The major types of heating elements are listed in *Table A.3*. In the following sections are described five types of resistance-heated electrical furnaces, followed by a description of gas-fired furnaces. Temperature control is the subject of the final section of this appendix.

There are other furnaces that could be used for the heat treatment of gemstones beside the resistance furnaces here described. Examples are radio-frequency induction furnaces, arc furnaces, and image furnaces of various types.

Table A.3 Electrical-resistance heating-element materials

Material	*Max. useful temp.* (°C)	*Comments*[a]
Chromium alloys		
Chromel C, Nichrome, Kanthal DT	1100	ONR
Kanthal A, Chromel A	1300	ONR
Precious metals		
Platinum	1400	ONR
Platinum–rhodium alloys	1500–1700	ONR
Other		
Silicon carbide (Globar, Crystolon)	1500	ON
Molybdenum disilicide (Kanthal Super)	1700	ON
Molybdenum	1800	NR
Lanthanum chromite (Keramax)	1800	O
Thoria, stabilized	2000	ONR, shock
Zirconia, stabilized	2800	ONR, shock
Tungsten	2800	NR
Graphite	3000	NR

[a] Usable in oxidizing, O, neutral, N, or reducing, R, atmospheres; 'shock' indicates particularly poor resistance to thermal shock, that is to sudden temperature changes.

Although these may be used on gemstones if they happen to be available, they are not the furnaces of choice and are not further discussed herein.

The smaller of the resistance-heated furnaces, such as a small box furnace capable of reaching about 1100 °C, merely need to be plugged into a suitable wall outlet. Where larger and higher-temperature furnaces are involved, the electrical requirements should be carefully checked before ordering, since special high-current and high-voltage circuits may have to be installed. Some furnaces may also require cooling water and special, protective, atmosphere gases. Good ventilation is always advisable since toxic gases may be emitted.

The actual choice of the spacing of the heating elements, their number, size, etc., can be quite complex. Thus, in a wire-wound furnace the resistance of the winding is a function of the wire diameter and length; the total watts of power emitted also depend on these factors, as do the watts per unit surface area of the wire, which must be kept within reasonable limits to avoid hot spots and localized melting. Thus, while a certain winding at 500 °C may have an energy density of 20 watts per square inch, a value of 4 is more reasonable at 1000 °C. Details are given in various places[1, 9, 10].

The electrical voltage or current that operates the furnace may be set manually by a variable resistance or transformer. Control of the temperature via a thermocouple or pyrometer, as described at the end of this appendix, may be achieved simply by turning the power on and off with an electro-mechanical relay or an electronic silicon-controlled switch, which pulses the power at a variable rate and provides much better control and a longer heating-element lifetime. Better yet is the use of silicon-controlled rectifiers or saturable core reactors, which give a continuously variable output and are used in more sophisticated furnaces. Details are best obtained from the manufacturers of temperature-control equipment (*see* Appendix D).

It is generally true that furnace heating-elements have a longer operating lifetime if they are not turned off between use (except for extended periods, of course); they should not, however, be left at their maximum temperature unnecessarily, but turned down to perhaps one-half to two-thirds of the maximum temperature.

A Resistance Box or Muffle Furnace

A box furnace has a hot region approximately the shape of a box with sides of roughly equal dimensions. Commercial furnaces of this type are widely used in the laboratory and readily available with maximum operating temperatures of about 1000 or 1200 °C, depending on which of the first two items of *Table A.3* are used for the heating elements. An example is shown in *Figures A.1* and *A.2*. The heating elements have a limited life at the highest temperatures, but are readily replaced, one at a time, if damaged or burned out. They are usually available as flat, ceramic plates which are grooved; the spiral wire-elements lie in the grooves and frequently are covered with a ceramic cement for support and protection; an example of an uncoated element is shown in *Figure A.3*. Four such plates make up the top, bottom, and two sides of the box, as shown in *Figure A.2*, surrounded by the insulating firebrick of *Table A.2*, and an outer, metal shell. This shell also holds a

Figure A.1 An electrically powered box or muffle furnace capable of 1200 °C using panels, as in *Figure A.3*, with thermocouple temperature-measurement; the construction is shown in *Figure A.2*. Photograph by courtesy of Thermolyne, a Division of Sybron Corp

dial that sets the power level, and a simple temperature indicator activated by a thermocouple (*see* the last section of this appendix), which projects through the back wall into the heated space. Alternatively, there may be a temperature-selecting dial that operates the furnace from a thermocouple via a mechanical or electronic relay. Although usually only used in air, it would be possible to modify such a furnace to employ other atmospheres. The rather low temperature limit of such furnaces is offset by their relatively low cost and simplicity of installation, operation, and repair.

Precious-metal Pot and Tube Furnaces

Temperatures up to 1700 °C are available with platinum–rhodium alloy wire in horizontal or vertical tube furnaces, which are open at both ends, or in pot furnaces, which are open only at the upper end, as in *Figure A.4*. Ceramic furnace cores are used, usually made of alumina and with a spirally grooved outer-surface so that the resistance-heating wire can be wound in the groove, as shown in *Figure A.3*. A core has the advantage over a smooth tube that the winding is uniform and

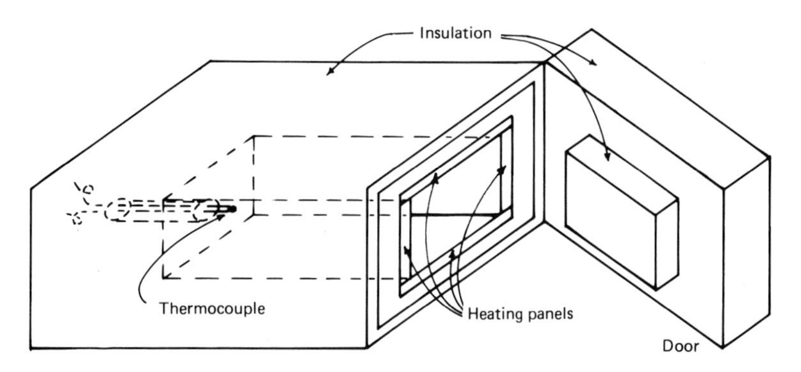

Figure A.2 The construction of a box or muffle furnace, as in *Figure A.1*

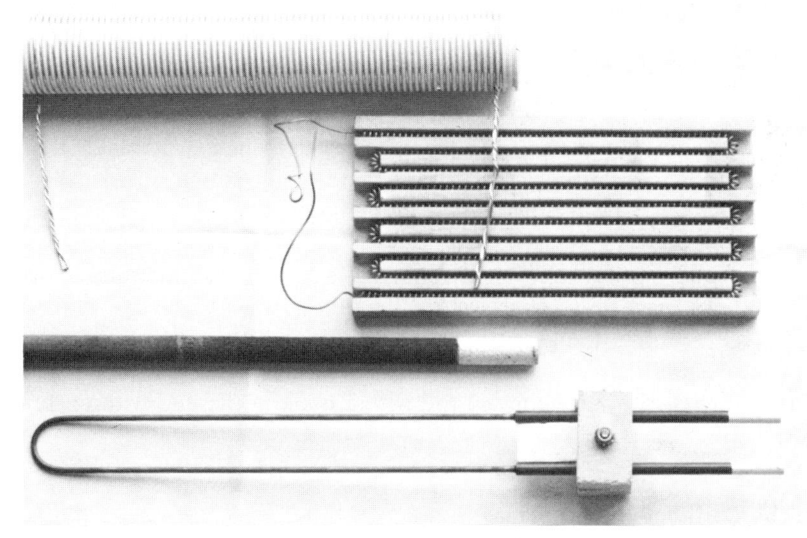

Figure A.3 Electrical heating elements. Top to bottom: platinum–rhodium wire-wound core for a tube furnace capable of 1700 °C; resistance-wire panel as used in the box or muffle furnace of *Figures A.1* and *A.2*; a silicon carbide element as used in the furnaces of *Figures A.5* and *A.7*; a molybdenum disilicide U-shaped heating element, as used in the furnaces of *Figures A.8* and *A.9*

that adjacent turns cannot touch each other. This could be a problem if a smooth tube is used because the wire expands on heating and so may become loose. A layer of furnace cement, such as one of the *Alundum* cements made by the Norton Co, is usually applied to cover the winding on the core. This helps to hold the wire in place, evens out the temperature, and protects the winding as well. Both core and cement must, of course, be electric insulators.

The ends of a core are usually first equipped with a single-turn loop of the wire with the two ends twisted together. This provides anchors while winding the core and also provides a multiple-strand termination of lower resistance for the heating wire, as shown in *Figure A.3*; because of this, only a little heat is liberated in these connections compared with the winding. As shown in *Figure A.4*, the core is usually

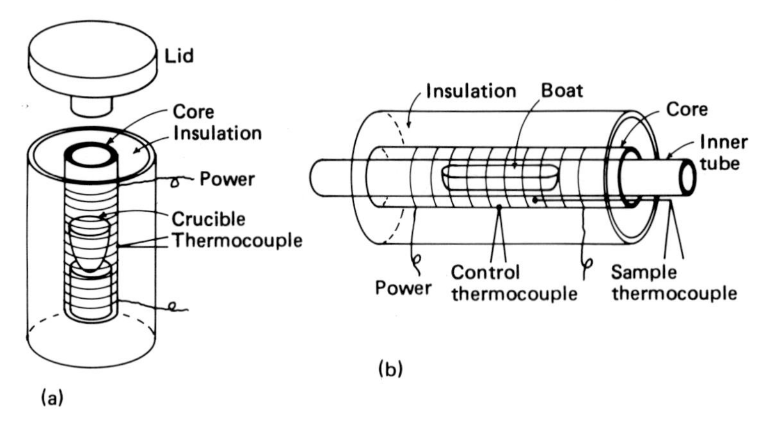

Figure A.4 The construction of (a) a pot furnace and (b) a tube furnace, both using a wire-wound core, as in *Figure A.3*, for the heating element

positioned in the center of a steel or aluminum shell filled with insulation, such as the alumina bubble material of *Table A.2*. A thermocouple touching the outside of the cement surrounding the tube is often used to control the temperature via an electronic power supply, while another thermocouple can be placed inside the tube in the hot region to measure the actual sample temperature; alternatively, the latter may be used to perform both functions or there may be two thermocouples inside the tube.

If an atmosphere other than air is required, a longer ceramic tube is usually placed inside the furnace core, as shown for the tube furnace in *Figure A.4*, so that any desired gas can now be passed over the sample, which is usually placed in a small, ceramic boat. If the material can tolerate the heat shock, the boat may be inserted or removed with a long metal rod with a hook at the end without turning off the furnace.

A Silicon Carbide Muffle Furnace

Sintered silicon carbide ceramics are used as rod or tube heating-elements at temperatures up to 1500 °C, particularly where large volumes need to be heated. These elements are sold under a variety of trade names, such as *Globar* by the Carborundum Co and *Crystolon* by the Norton Co. One such element is included in *Figure A.3*. These materials have the characteristic that their resistivity falls as the temperature rises, so that by applying a fixed voltage, only a small current flows at first, but increases rapidly as the temperature rises; special current-limiting control circuits must, accordingly, be used to avoid an electrical overload. The elements have lower resistance sections at each end so that heat is generated only in the central segment. Electrical characteristics vary from element to element and also change with time. Matched sets of elements are usually used; if one is damaged, all are then replaced.

A typical arrangement for a muffle furnace is shown in *Figure A.5,* in which the sample is located in the 'muffle', a box with a door at the front, also usually made of sintered silicon carbide. Compare this with the illustration of a solid-fuel fired muffle-furnace from Agricola's 1556 book[12], shown in *Figure A.6.* Above and

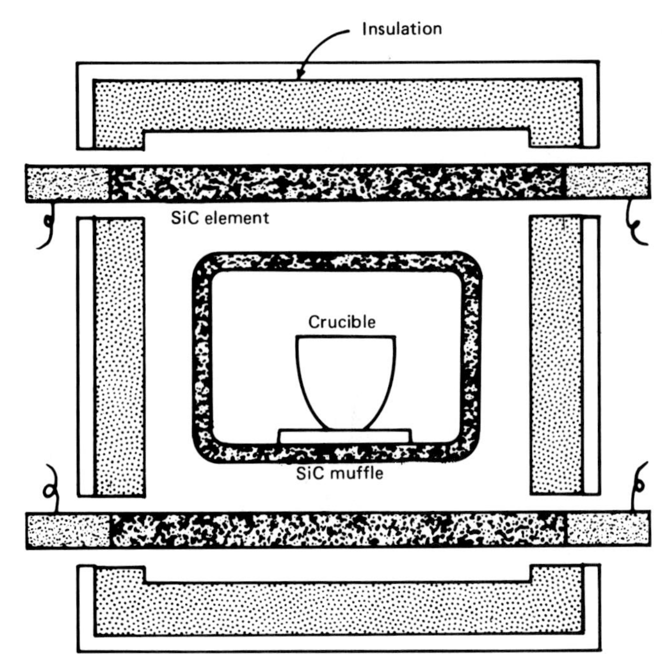

Figure A.5 The construction of a silicon carbide muffle furnace, as in *Figure A.7*

Figure A.6 A solid-fuel fired muffle-furnace from Agricola's 1556 book on metallurgy

Figure A.7 A muffle furnace, capable of 1500 °C, using silicon carbide elements, as in *Figures A.3* and *A.5*. Photograph by courtesy of Harper Furnaces and Kilns

below the muffle in *Figure A.5* are the horizontal heating-elements that pass through and are supported at their cooler ends by the firebricks constituting the insulating walls of the furnace, in turn surrounded by an outer metal shell, as shown in *Figures A.5* and *A.7*. A makeshift arrangement sometimes seen in the laboratory consists of firebricks merely stacked together into a large block with a space at the center and holes cut to accept the silicon carbide elements. Special aluminum connectors need to be used at the ends of the elements and are available from the manufacturers.

A Molybdenum Disilicide Box-furnace

The sintered, ceramic molybdenum disilicide heating-elements made by the Kanthal Corp under the name of *Kanthal Super* provide a convenient way to reach temperatures up to 1700 °C. They are made in the form of U-shaped elements, one

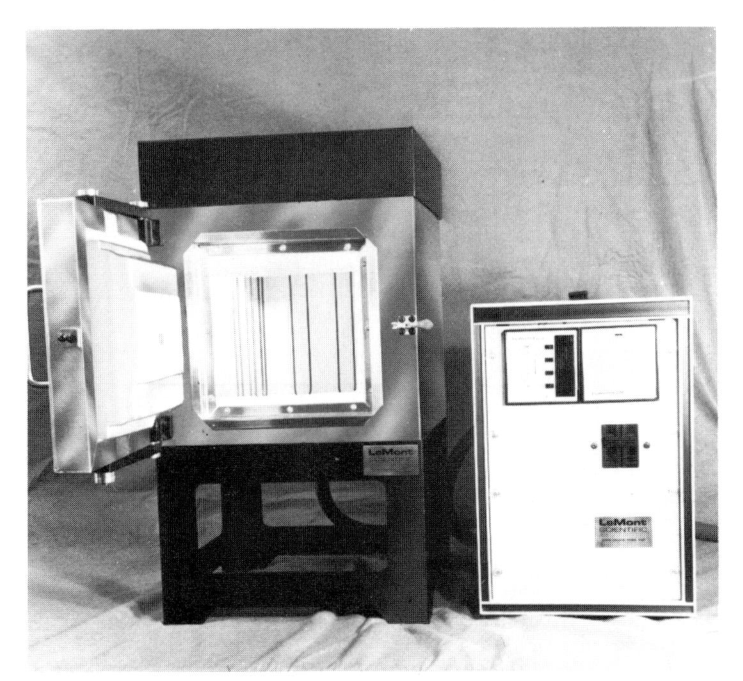

Figure A.8 A box furnace, capable of 1700 °C, using molybdenum disilicide elements, as in *Figure A.3.* Photograph by courtesy of Le Mont Scientific

Figure A.9 A vertical tube-furnace, capable of 1700 °C, using molybdenum disilicide elements, as in *Figure A.3.* Photograph by courtesy of C M Manufacturing and Machine Co

Figure A.10 A graphite-tube furnace capable of 2750 °C, equipped with a strip-chart instrument for recording the temperature. Photograph by courtesy of Centorr Associates

being included in *Figure A.3*. These elements are usually used vertically, penetrating through slits in the top wall of a box furnace to project downward, and ranged along the inner sides of the box, as shown in *Figure A.8*, or in a vertical-tube furnace, as in *Figure A.9*.

The resistance of molybdenum disilicide heating-elements increases rapidly with temperature, so that a relatively high voltage can be applied from the start. This produces rapid heating with an automatic reduction in power as the working temperature is reached. Heating from room temperature to 1600 °C can be achieved in less than 15 minutes! Distortion of the elements occurs if the furnace is opened at very high temperatures and this leads to a shortened life; damaged elements can be replaced individually. Here, too, special aluminum connectors are used to connect the heating elements. Although a hydrogen atmosphere is not recommended, a mixture of hydrogen and water is usable with an occasional oxidation run in air. Separate atmospheric control can be achieved in special configurations.

A Graphite-tube Furnace

Carbon in the form of graphite (as well as tungsten at times) is used as the heating element in furnaces that can reach 3000 °C. *Figure A.10* illustrates a horizontal, graphite-tube furnace. The arrangement is similar to that of the tube furnace illustrated in *Figure A.4,* except that the wire-wound core is replaced by a graphite tube with a current-carrying graphite clamp at each end. To increase the resistance, slots are often cut into the graphite tube so that the current must follow a spiral or zig-zag path. The carbon or graphite powder of *Table A.2* is used as insulation, and a protective atmosphere (often employing the relatively safe 'forming gas' discussed in Chapter 3) is used to prevent oxidation of the tube and insulation. Since some hydrocarbon gases can be produced from hydrogen in the protective atmosphere and some cyanogen from nitrogen, good ventilation must be provided for the spent protective gas.

If an oxidizing atmosphere is required for the sample, an inner, ceramic tube made of high-purity alumina can be used for temperatures up to 1900 °C, and the stabilized zirconia of *Table A.3* is used for higher temperatures, in an arrangement somewhat similar to that shown for the tube furnace of *Figure A.4.* There is, however, a serious explosion hazard if the ceramic tube were to crack during operation and permit mixing of hydrogen and oxygen, for example. Since thermocouples are not available for the highest temperatures (except for the tungsten–rhenium ones, only usable in a non-oxidizing atmosphere, as discussed below), radiation pyrometers are used for temperature control, as also discussed below.

Gaseous Fuel Furnaces

It is possible to achieve quite high temperatures by using various gaseous fuels, such as coal gas, propane, hydrogen, acetylene, or even vaporized gasoline in combination with compressed air or oxygen. The maximum temperature available

Table A.4 Approximate maximum temperatures available from some gaseous fuels

Gas combination	*Temperature*[a] (°C)
Gas and air	1800
Hydrogen and air	2000
Gas and oxygen	2300
Hydrogen and oxygen	2500
Acetylene and oxygen	3000

[a] The actual temperature reached depends on the gas flow and the thermal insulation.

in some flames is given in *Table A.4.* It should be noted that there is considerable variability here, depending on the rate of the gas flow as well as on the arrangement used and the efficiency of the thermal insulation. Temperature control may be derived from a thermocouple for lower temperatures or, more commonly, from a radiation pyrometer, as seen on a tripod in front of the gas-fired furnace shown in *Figure A.11*.

In a relatively simple arrangement frequently used in the absence of technological sophistication, a steel drum is lined with a thick layer of fireclay (or even ordinary clay) and a flame is injected either from the bottom or tangentially from the side, as shown in *Figure A.12*. The torch is aimed to one side, away from the

Figure A.11 A gas–oxygen fueled furnace, capable of 2100 °C, using a radiation pyrometer for temperature control. Photograph by courtesy of Le Mont Scientific

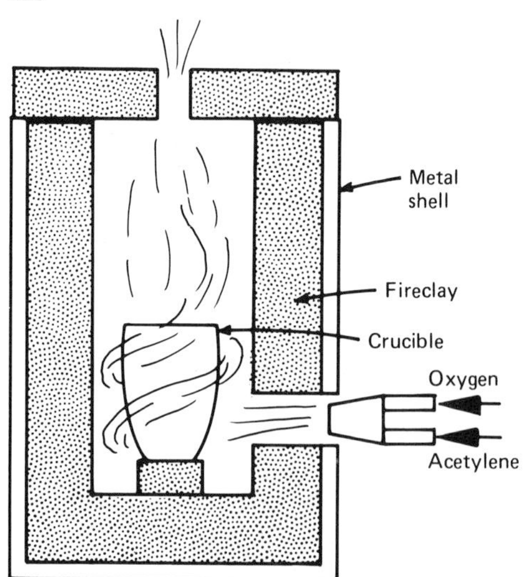

Figure A.12 Construction of a simple gaseous-fuel fired furnace, as in *Figure A.13*

crucible, with the flame swirling around the crucible; this provides an even temperature without hot spots. *Figure A.13* shows such an arrangement, photographed in Sri Lanka. The widely available oxygen–acetylene combination used for welding is frequently used. If the material being treated is in an open crucible, then

Figure A.13 A simple gaseous-fuel fired furnace, as in *Figure A.12*, used in Sri Lanka. Photograph by courtesy of the Gemological Institute of America

it is exposed to the atmosphere of the flame combustion products. Reducing conditions can be achieved simply by cutting back on the oxygen flow to achieve a sooty carbon- and carbon monoxide-rich flame and atmosphere:

$$C_2H_2 + O_2 \rightarrow H_2O + CO + C \qquad (A.1)$$

while an oxygen-rich flame produces an oxidizing atmosphere:

$$C_2H_2 + 3O_2 \rightarrow H_2O + 2CO_2 + \tfrac{1}{2}O_2 \qquad (A.2)$$

Reducing and oxidizing conditions can similarly be obtained with other gas combinations.

A serious drawback of such a simple arrangement is the difficulty of controlling the temperature, with a puddle of molten gemstones reported to be a frequent product.

Temperature Measurement and Control

Mercury-in-glass thermometers are available with a temperature range extending to over 500 °C, but they are very fragile and cannot take much thermal shock. More useful in the range up to 500 °C are bimetallic dial-reading thermometers, as shown in *Figure A.14,* which are usually made of stainless steel.

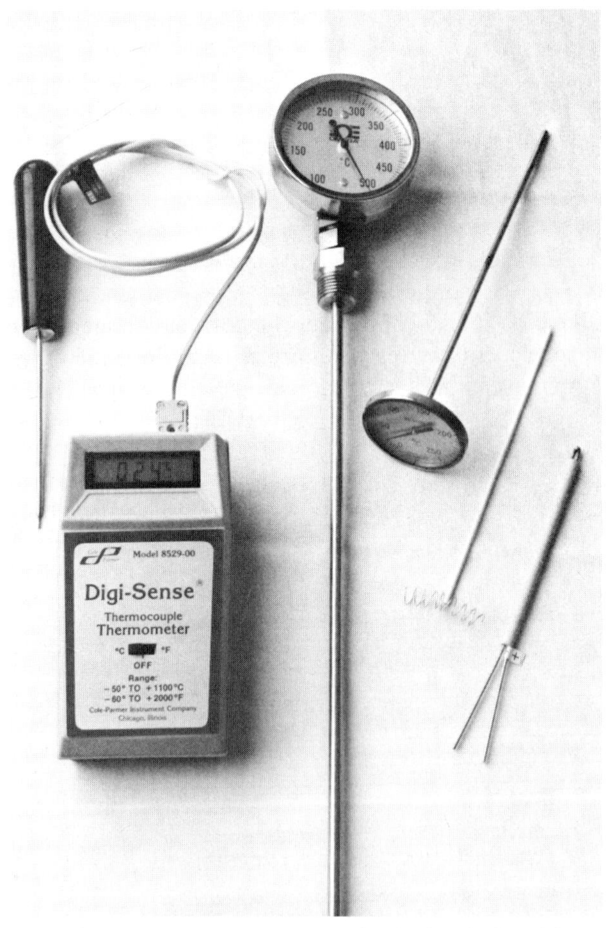

Figure A.14 Some temperature measuring devices. Left to right: a battery-operated digital thermocouple unit, capable of 1100 °C; two bimetallic, stainless steel stem thermometers, capable of 500 and 250 °C, respectively; a Type B platinum–rhodium thermocouple, capable of 1700 °C; and a Type K chromel–alumel thermocouple, capable of 1250 °C

The temperature-sensing and -controlling elements most commonly used in furnaces are thermocouples. If two different metals are joined and the junction is heated, then a small voltage appears. To measure this voltage a second junction, the 'reference' or 'cold' junction, is necessary so that the two wires leading to the measurement instrument are made of the same material, as shown in *Figure A.15*.

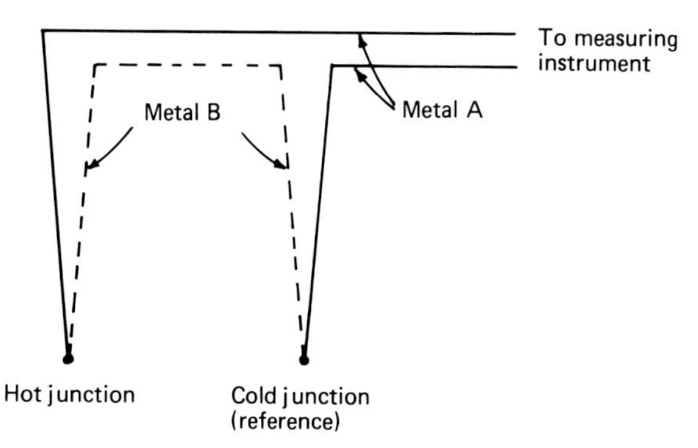

Figure A.15 The basis of thermocouple temperature-measurement devices

The voltage produced is then the difference between the voltages produced at the two temperatures. The reference junction may actually be kept in a thermos jug containing ice and water at 0 °C, but more usually a circuit is built into the measuring instrument which electronically provides the correct voltage to compensate for the cold junction; special connecting wires are also employed. Examples of commonly used thermocouple combinations are given in *Table A.5*; the higher the output voltage, the simpler the associated circuitry and/or the more accurate the temperature reading.

The measurement instrument may be a low-cost needle-indicating dial meter with a simple, built-in cold junction. Much more accurate are sophisticated, digital,

Table A.5 Characteristics of thermocouples

Type[a]	*Combination of metals or alloys*	*Output at 900 °C* (mV)	*Temperature limit* (°C)
T	Copper–constantan	(20.9[b])	400
J	Iron–constantan	(21.9[b])	760
E	Chromel–constantan	68.8	900
K	Chromel–alumel	37.3	1250
S	Platinum–platinum 10% rhodium	8.4	1450
R	Platinum–platinum 13% rhodium	9.2	1450
B	Platinum 6% rhodium–platinum 30% rhodium	4.0	1700
G	Tungsten–tungsten 26% rhenium	12.3	2300[c]
C	Tungsten 5% rhenium–tungsten 26% rhenium	16.4	2300[c]
D	Tungsten 3% rhenium–tungsten 25% rhenium	15.1	2300[c]

[a] ANSI (American National Standards Institute) symbols, except for G, C, and D.
[b] Output at 400 °C.
[c] Non-oxidizing atmosphere only.

electronic devices, which may be portable, as is the battery-operated unit shown in *Figure A.14*; with sophisticated cold-junction compensation these may give readings accurate to 1–3 °C. Finally, such an electronic unit may also incorporate sophisticated, multiple-action, temperature-control circuitry which sends a reference signal to a power supply that controls the energy flow to the furnace. A strip-chart recorder, as shown in *Figure A.10,* may complete such an arrangement, providing a record of the actual variation of temperature-with-time within the furnace.

Thermocouples can deteriorate and give erroneous readings if either the metal or the supporting ceramic tube becomes contaminated with one of a large variety of substances. Accordingly, they are often surrounded by a protective metal tube, as in the probe attached to the unit shown in *Figure A.14.*

Pyrometers can be used for the measurement and control of temperatures from about 900 °C up to beyond the limit of thermocouples. It is well known that the hotter an object, the more light it emits, with a color that depends on the temperature; hence, red-hot, orange-hot, white-hot, and so on. An experienced furnace operator can estimate a temperature above 700 °C to better than 50 °C merely by judging the intensity and the color of the light. Pyrometers are instruments that perform this same function, but more precisely. In an 'optical pyrometer' the brightness seen through a red filter in a small viewing telescope is matched by the brightness of the filament in a light bulb, with the variable filament current calibrated to give a direct reading of the temperature.

In 'radiation pyrometers' the eye is not involved; instead the light (radiation) from the hot region is focused onto a 'photodetector' in the pyrometer. This produces a voltage which can give a temperature indication on a digital instrument, as in *Figure A.16*, or which can control an electrical power supply or a gas-flow

Figure A.16 A battery-operated, digital radiation pyrometer. Photograph by courtesy of Omega Engineering Inc

system, as in *Figure A.11*. To function properly, a radiation pyrometer must have a clear view of some part of the hottest region of the furnace, but it does not need to be in contact with it. If there is a partial obscuration of the view, there are even 'two-color pyrometers' available, which can function under such difficult conditions.

It must be kept in mind that in the absence of calibration, a pyrometer does not give an absolute temperature value. For a true value, one would need to know the *emissivity*, which is a measure of the efficiency of light emission of the surface being viewed; this is a parameter that is rarely determined. In actual practice it is assumed that the emissivity is unity, i.e. that the furnace is acting as an ideal 'black body', and the value obtained is then the 'uncorrected pyrometric temperature'. For most purposes this is perfectly adequate, since consistency in the control of the temperature is more important than knowing the actual value of the temperature. One could always calibrate at 2050 °C by melting some colorless sapphire or highly purified alumina or some other pure substance, usually a metal[6] of precisely known melting point, not too far from the temperature being measured. The accuracy of pyrometric temperatures is typically 5 to 10 °C.

In this field, too, books[4, 6], as well as manufacturers' literature[11] should be consulted for further details (reference 11 contains an excellent outline of thermocouple theory and practice, as well as reference tables, pp. T1–T79).

References

1. V. Pashkis and J. Persson, *Industrial Electric Furnaces and Appliances,* 2nd Edn, Interscience Publishers, New York (1960)
2. M. W. Thring, *The Science of Flames and Furnaces,* Wiley, New York (1952)
3. P. L. Start and M. W. Thring, Design of laboratory furnaces, *J. Sci. Instr.,* **37,** 17 (1960)
4. P. A. Kinzie, *Thermocouple Temperature Measurement,* Wiley, New York (1973)
5. W. D. Kingery, H. K. Bowen, and D. R. Uhlmann, *Introduction to Ceramics,* 2nd Edn, Wiley, New York (1976)
6. I. E. Campbell and E. M. Sherwood (Eds), *High Temperature Materials and Technology,* Wiley, New York (1967); in particular: J. M. Smith, Selection of materials, pp. 131–151; S. W. Bradstreet, Oxide ceramics, pp. 235–303; H. D. Sheets and J. D. Sullivan, Commercial oxide ceramics, pp. 304–311; G. H. Kessler, Thermal insulation, pp. 693–714; and J. I. Slaughter and J. L. Margrave, Temperature measurement, pp. 717–792
7. A. Smakula, *Einkristalle,* pp. 180–213, Springer Verlag, Berlin (1962)
8. No author, *Laboratory Electric Furnace Construction using Cores, Tubes and Muffles,* Norton Co, Worcester, MA 01606 (1976)
9. No author, *The Kanthal Handbook,* Kanthal Corp, Bethel, CT 06801 (1975)
10. No author, *Kanthal Super and its Uses in Furnaces,* Kanthal Corp, Bethel, CT 06801 (1977)
11. No author, *Omega Temperature Measurement Handbook, 1984,* Omega Engineering Inc, Stamford, CT 06907 (1984)
12. G. Agricola, *De Re Metallica,* Basel (1556)

Appendix B

More on Irradiation

This appendix is intended to supplement the descriptions in Chapter 4, which should be read first.

An excellent place to find background material on radioactivity, radiation, and irradiation in general is Glasstone's *Sourcebook on Atomic Energy*[1]. Then there are books on radiation physics, such as that of Lapp and Andrews[2], on radioactivity, such as that by Coombe[3], on radiation sources, such as that by Charlesby[4], and on radiation protection, such as the ones by Shapiro[5] and Morgan and Turner[6]. Books such as the last two, as well as Sax's volume[7], are also essential reading for anyone working with radioactive materials, for their own protection as well as that of others.

Radiation and Irradiation Units

The units used for measuring the amount of a radioactive material are the curie (abbreviated Ci) and the becquerel (abbreviated Bq). These are defined in terms of the number of disintegrations per second (d/s); the millicurie and microcurie definitions are also given*:

1 Ci produces 37 000 000 000 d/s
1 mCi produces 37 000 000 d/s
1 μCi produces 37 000 d/s
1 Bq produces 1 d/s

Accordingly, 1 Ci = 3.7×10^{10} Bq.

The unit used for exposure to gamma radiation and X-rays is the roentgen (abbreviated R, but r is also used at times). This is defined in terms of the number of ionizations the radiation produces in air: one R is that amount of material which produces a quantity of ions corresponding to 1 electrostatic unit of charge in 1 cc of dry air at standard temperature and pressure.

* The unit prefixes most commonly used, their abbreviations, and their multipliers are: giga, G, 10^9; mega, M, 10^6; kilo, k, 10^3; centi, c, 10^{-2}; milli, m, 10^{-3}; micro, μ, 10^{-6}; and nano, n, 10^{-9}.

When ionizing radiation of any type interacts with matter, the unit used for the absorbed dose is the rad (no abbreviation). One rad is defined as yielding the release of an energy equal to 100 ergs per gram of absorbing matter. A recently introduced unit is the Gray (abbreviated Gy) with 1 Gy = 100 rad.

Very frequently the absorbed dose equivalent in a biological system (e.g. man) is of interest, when the rad is replaced by the rem ('roentgen-equivalent-man'). Different types of radiation have different abilities to do damage; one rem is that amount of ionizing radiation which does the same amount of damage in man as 1 R of 200 000 V X-rays. Although there are subtle differences, one can assume that for X-rays, gamma rays, and electrons, 1 R = 1 rad = 1 rem to a first approximation. For slow neutrons 1 rad produces about 2.5 times the damage of 1 rad of X-rays, etc., so that here the number of rems is about two-and-a-half times the number of rads; for fast neutrons and alpha particles the number of rems is about ten times the number of rads. A recently introduced unit is the Sievert (abbreviated Sv), with 1Sv = 100 rem.

Dose rates for any of the dose units are most commonly given per hour or per year, as in rad/h or rad/yr. The total dose absorbed is then given by the number of dose units per unit of time multiplied by the total time of exposure. Consider the exposure of a person for 6 h to 1 rad per h of gamma rays and 2 rads per h of fast neutrons. The exposure rate is $1 + (10 \times 2) = 21$ rems/h, and the total exposure is $21 \times 6 = 126$ rems, a dose sufficiently large to produce prompt symptoms of radiation sickness. The same amount of radiation exposure over a period of one year would produce essentially no symptoms. As a reference figure, the background exposure from cosmic rays and natural radioactivity in an average environment is about 0.012 millirems per h or 0.1 rems per year.

It should be noted that the rad is a unit that is independent of the material absorbing the radiation, while the other units vary depending on what fraction of the radiation is absorbed.

Atomic Structure and Radiation

The simplest atom, that of hydrogen, consists of a nucleus having an electron in orbit around it. The nucleus is a proton, a particle of unit positive charge and unit atomic mass as shown in *Figure B.1*. The electron also shown there has a unit negative charge and a tiny mass, only about 1/1823 the mass of the proton. The atomic mass, $M = 1$, is thus dominated by the nucleus, and the atomic number, $Z =$ 1 in element number one, is given by the charge on the nucleus. An atom of hydrogen, shown in *Figure B.1,* can be designated as ${}^{1}_{1}H$, where the upper number is M and the lower is Z. In heavy hydrogen, there is a neutron also present in the nucleus. Since the neutron has no charge and a mass of one, as shown in *Figure B.1,* this isotope of hydrogen, designated as ${}^{2}_{1}H$ and also shown in *Figure B.1,* can also be written ${}^{2}_{1}D$, for deuterium is its alternative name. The lower number thus specifies the element (1 for hydrogen), the upper the mass; deuterium is the heavy isotope of hydrogen.

The next element of interest is helium, ${}^{4}_{2}He$. The lower number specifies the charge on the nucleus, thus indicating the presence of two protons in element

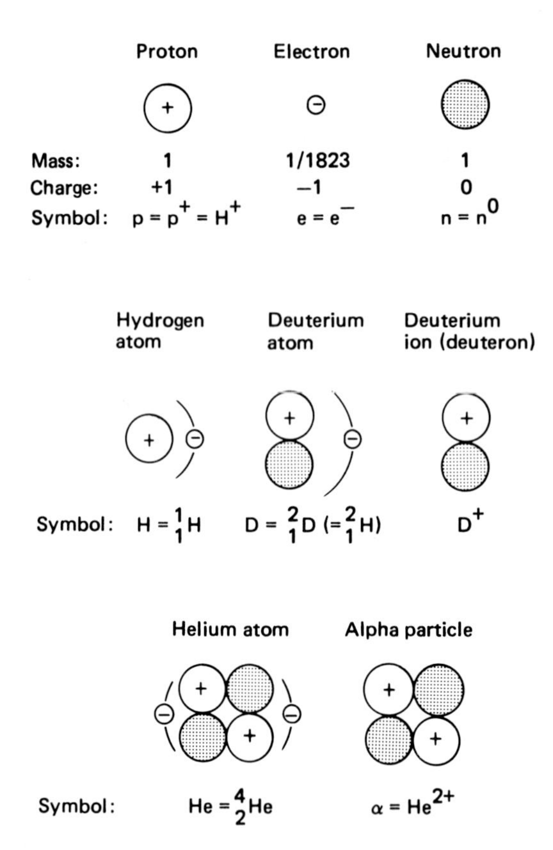

Figure B.1 Particles involved in irradiation

number two. The upper number gives the nuclear mass as four, so that, in addition to the two protons, two neutrons are indicated as being present. Since the atom as a whole is neutral, there are also two negatively charged electrons in orbit, as shown in *Figure B.1*.

The symbol for ordinary cobalt, $^{59}_{27}Co$, shows that the nucleus contains a positive charge of 27 units, i.e. 27 protons, this being element number 27. Subtracting the mass of the protons from the nuclear mass of 59, 59 − 27, leaves 32 neutrons to complete the nucleus; 27 electrons in orbit complete the neutral atom as found in metallic cobalt. The 27 electrons in orbit are arranged into shells; the innermost first shell contains two electrons; the second shell eight, the third shell 15 and the fourth shell the remaining two. The two innermost shells of this electronic arrangement are completely filled and contain only paired electrons, while the outermost electrons are not all paired and some may be donated to other atoms to form ions, as in ionic bonding, or shared, as in covalent or metallic bonding.

Most solid substances when heated in a vacuum emit free electrons from their outermost shells; these negatively charged electrons can be accelerated in linear accelerators or betatrons to produce the energetic beta-particles described in Chapter 4 and below. Energetic electrons are also emitted by some radioactive nuclei.

Since the 'electron volt' energy unit is defined in terms of the energy of one electron accelerated by a potential of 1 V, the electrons produced in a linear accelerator, as in *Figure B.2(a)* have an energy equal to the voltage applied.

If an energetic electron interacts with an atom of cobalt, it can displace one of the two innermost electrons out of position, as shown in *Figure B.3(a)*. When one of the outermost electrons falls back down to replace the missing innermost electron, there remains an excess energy equal to the difference in energy between these two electron states, namely 6.9 keV; this is emitted as 'characteristic' electromagnetic radiation and is, in fact, an X-ray having an energy of 6.9 keV, which corresponds to a wavelength of 0.179 nm. Other electrons may also be involved in this re-filling so that there is produced a set of X-ray energies that is characteristic of the metal used. A second process, producing 'Bremsstrahlung' X-rays, involves the radiation of a wide range of energies when an electron is deflected by passing close to an atomic nucleus, as shown in *Figure B.3(b)*. These are the processes which occur in an X-ray tube, as in *Figure B.2(b)*, in which other target metals commonly used include tungsten, molybdenum, copper, iron, and so on.

If a thin stream of hydrogen gas is permitted to enter a vacuum and is ionized, that is if each atom has its single electron removed, the positively charged ions produced can then be accelerated in an electric field. The result is a stream of energetic hydrogen nuclei or protons, as in the chemical type equations:

$$H \rightarrow H^+ + e^- \quad (B.1)$$

or

$$H \rightarrow p^+ + e^- \quad (B.2)$$

where e^- is the negatively charged electron and p^+ is a single, positively charged proton.

If the same is done with helium gas, the result of a double ionization is a doubly charged helium nucleus, also called an alpha particle when emitted from a radioactive nucleus; it too is shown in *Figure B.1*:

$$He \rightarrow He^{2+} + 2e^- \quad (B.3)$$

Consider now some metallic cobalt which is exposed to neutrons in a nuclear reactor. A neutron, $^{1}_{0}n$, can enter the cobalt nucleus and convert the $^{59}_{27}Co$ into $^{60}_{27}Co$, also known as cobalt 60, written this time as a nuclear equation:

$$^{59}_{27}Co + ^{1}_{0}n \rightarrow ^{60}_{27}Co \quad (B.4)$$

Cobalt 60 is unstable and decomposes; this process can be viewed as one of the neutrons breaking up into a proton plus an electron:

$$n \rightarrow p^+ + e^- \quad (B.5)$$

leading to the effective change:

$$^{60}_{27}Co \rightarrow ^{60}_{28}Ni + e^- + \text{energy} \quad (B.6)$$

The ejected electron or beta particle, e^-, has an energy up to 319 keV; the energy given off in equation (B.6) consists of equal amounts of gamma rays, of

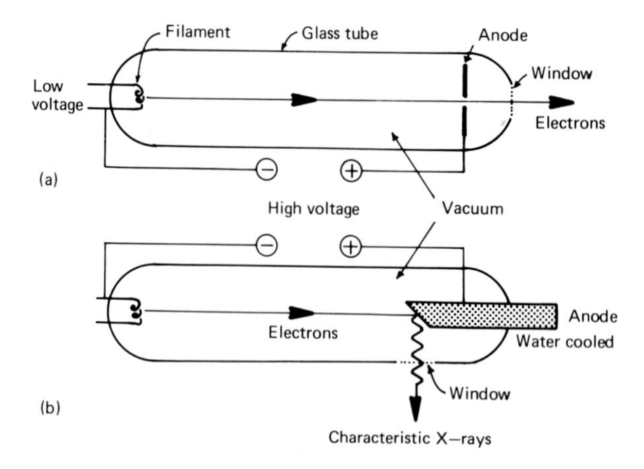

Figure B.2 (a) The acceleration of electrons is (b) also used to produce X-rays

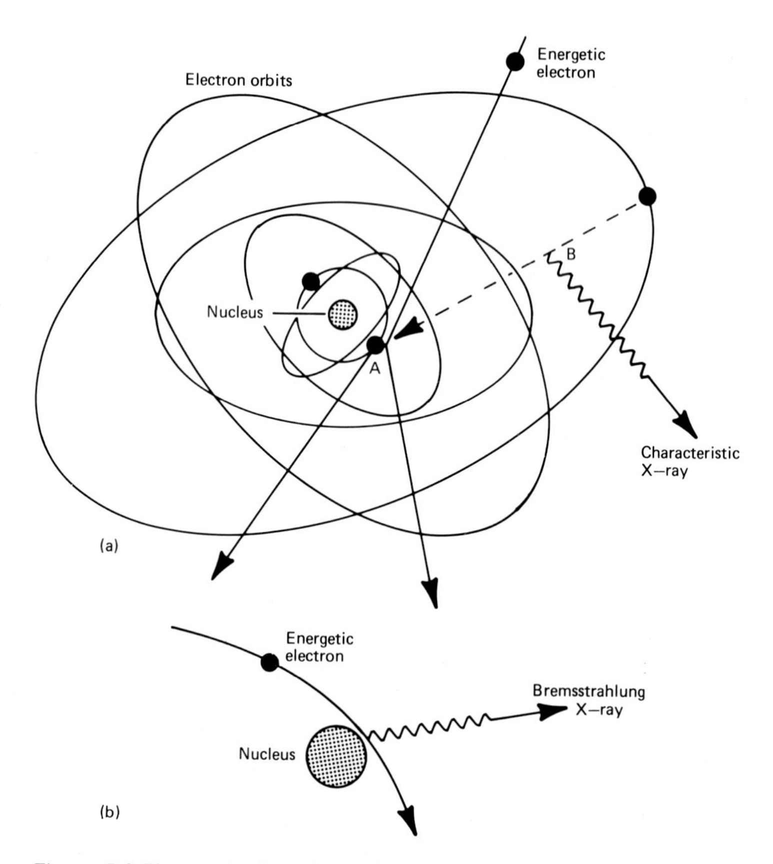

Figure B.3 The production of (a) characteristic X-rays by a high-energy electron which ejects a low-lying electron and of (b) Bremsstrahlung X-rays by the deflection of an electron passing close to a nucleus

energies 1.17 MeV and 1.33 MeV. The presence of the beta particles can generally be ignored, since they are usually absorbed by the sample container. These are the processes that occur in the gamma cell of *Figure 4.2*.

Another example of the production of gamma rays is from cesium-137, a radioactive isotope produced during the fission of uranium in a nuclear reactor and isolated during the chemical processing of spent nuclear-reactor fuel. This atom is unstable and emits a beta particle of energy 0.51 MeV, thereby transforming to barium-137 with a half-life (*see below*) of 30 years:

$$^{137}_{55}\mathrm{Cs} \rightarrow \mathrm{e}^- + {}^{137}_{56}\mathrm{Ba} \qquad \text{(B.7)}$$

The barium itself is an unstable isotope having an excess of energy, which is emitted as a gamma ray of energy 0.66 MeV with a half-life of 2.6 minutes.

Radium, Ra, is an element which can be refined from uranium ore, where it is produced by the radioactive disintegration of the uranium itself via a complex decay sequence. The unstable radium emits an alpha particle, the nucleus of a helium atom, with energy 4.8 MeV, leaving behind an atom of radon, Rn, as follows:

$$^{226}_{88}\mathrm{Ra} \rightarrow {}^{4}_{2}\mathrm{He} + {}^{222}_{86}\mathrm{Rn} \qquad \text{(B.8)}$$

The half-life of this decomposition is 1602 years. Additional gamma rays are produced by the radon, which has a half-life of only 3.8 days and the subsequent decomposition products $^{218}_{84}\mathrm{Po}$ and $^{214}_{84}\mathrm{Po}$, which have even shorter half lives.

In a nuclear reactor many reactions occur, but a single example will suffice. Consider one atom of the uranium isotope $^{235}_{92}\mathrm{U}$ reacting with a neutron to give an unstable nucleus which then splits apart:

$$^{235}_{92}\mathrm{U} + {}^{1}_{0}\mathrm{n} \rightarrow {}^{98}_{40}\mathrm{Zr} + {}^{135}_{52}\mathrm{Te} + 3\,{}^{1}_{0}\mathrm{n} \qquad \text{(B.9)}$$

Considerable energy is released in this process. Since one neutron produces several, this leads to the possibility of a chain reaction which can be run-away, as in a nuclear explosion, or controlled, as in a nuclear reactor used for power generation. Exposure to neutrons is usually performed in research reactors which have special access facilities, in which samples can be exposed to the neutrons.

Effects of Radiation on Matter[6]

There are three major effects of significance to gemstones when radiation interacts with matter. First, the radiation can enter the nucleus and produce nuclear reactions, as has been described above; this can happen with neutrons, energetic, heavy, positive particles, such as protons and deuterons, and with very energetic electrons (but not at energies of a few MeV, typically used for electron irradiation). One consequence of using any of the very energetic radiations is that radioactivity can be induced, as has been discussed in Chapter 4.

A second effect of high-energy irradiation is to produce drastic displacements of atoms aways from their regular position in a crystal. This has happened in a number of uranium- and thorium-containing minerals over geological periods. The outstanding example is zircon, and the consequence is the 'metamict' state, where sufficient disorder may be introduced to produce an amorphous or glass-like stone having an unusually low density. This is discussed in more detail in Chapter 7 under zircon.

The third effect is a displacement of the outermost electrons in atoms. This is the effect that can lead to valence-state changes and the formation of color centers, as has been described in Chapter 4. The ionization associated with the removal of outer electrons in biological materials can lead to short-term somatic effects, such as radiation sickness, and longer-term effects, such as the induction of carcinomas as well as mutagenic changes in the hereditary genes.

The relative effect on elemental germanium has been estimated[8] for typical intense-irradiation sources, using an exposure of one day each, in terms of the number of atoms displaced per incident particle, as shown in *Table B.1*. It can be seen that the heavier particles are much more efficient at causing effects, but at the expense of inducing radioactivity, as described previously.

Table B.1 Effect of different irradiations on germanium (modified after Billington and Crawford[8])

Irradiation	*Mass* (atomic mass units)	*No. of atoms displaced per incident particle*	*Exposure* (flux per day)	*Total effect* (displacements in one day)
Gamma rays (Co-60)	0	6×10^{-4}	10^{18}	6×10^{14}
Electrons (1.5 MeV)	1/1823	3×10^{-2}	10^{20}	3×10^{18}
Neutrons (1 MeV)	1	2.5	2×10^{19}	5×10^{19}
Deuterons (9.5 MeV)	2	8	10^{18}	8×10^{18}
Alpha particles (5.2 MeV)	4	43	10^{18}	4.3×10^{19}

Table B.2 The range in water of some particles

Particle	*Energy* (MeV)	*Mass* (atomic mass units)	*Velocity* (cm/sec)	*Range* (μm)
Electron	1	1/1823	2.8×10^{10}	4300
Proton	1	1	1.4×10^{9}	23
Alpha particle	1	4	0.7×10^{9}	7.2

An idea of the range, the limiting distance which particles can travel, is given in *Table B.2* as the ranges in water of three particles all having energies of 1 MeV. For the same energy, a lighter particle has a much higher velocity and a much higher range; the heavier particles interact much more intensively, as shown in *Table B.1,* and accordingly have a much shorter range. In substances other than the water of *Table B.2,* the range can be taken to vary inversely with the density in proportion to the range in water: in a material of density 2, the range is half that in water, and so on. It should be made clear that these ranges, such as the 4300 μm (4.3 mm) range of electrons in water, refer to the maximum range; the majority of the particles do not make it that far.

For controlling radiation, the type and quantity of shielding required varies greatly with the circumstances. Alpha particles are largely stopped by a sheet of paper, a rubber glove, or even by the layer of dead cells on the surface of human skin. Beta particles are much more penetrating, but are usually adequately shielded by 1 cm of plastic or by an aluminum sheet; since Bremsstrahlung X-rays can be generated in this process, a mm or so of lead is usually placed beyond the plastic or aluminum to absorb these secondary X-rays.

Lead is the customary absorber for X-rays and gamma rays, with a few mm generally being sufficient for X-rays, but 20 cm or more being required for high-energy gamma rays. Where bulk is no problem, steel, water, and concrete can be used. For gamma rays, one unit thickness of lead is approximately equivalent to two of iron, six of concrete, and twelve of water. Neutrons are shielded with hydrogen-containing materials such as water, wax, or plastics; boron in the form of boric acid is often added. The thickness of a shielding material is sometimes given as the 'tenth value thickness', this being the thickness that reduces the radiation intensity to one-tenth its original value. Twice this thickness reduces the radiation intensity to one hundredth, three times to one thousandth, and so on.

For small quantities of radioactive materials, such as a strongly radioactive gemstone, the best container is a lead box with an overlapping cap, with walls perhaps 2 cm thick and having a standard, radioactive-hazard label printed in magenta on yellow, as shown in *Figure 4.9*.

Particle Accelerators

In the preceding section and in *Figure B.2(a)* is described the acceleration of electrons by a high voltage in a linear accelerator. The problem here is to obtain an electric potential of 1 000 000 V plus for application to the accelerator. One

Figure B.4 Electron irradiation area adjacent to an electron accelerator (*see also Figures 4.4* and *4.5*). Photograph by courtesy IRT Corp

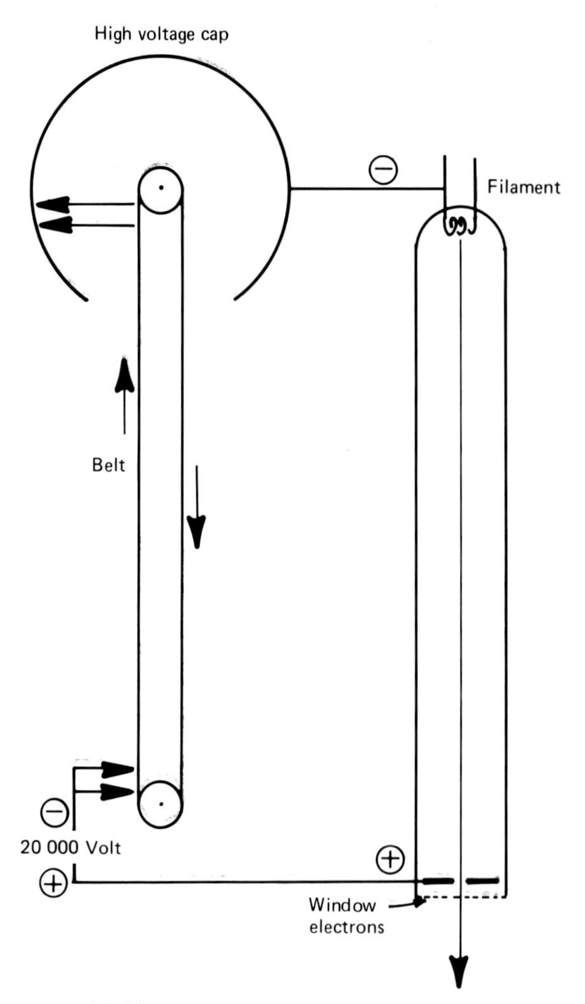

Figure B.5 Diagram of a Van de Graaff high-voltage generator powering a linear electron-accelerator

technique uses electronic voltage-doubler circuits, also known as voltage multipliers or cascade rectifiers; such a unit, with the capability of reaching 1.5 MeV at 25 mA, is shown in *Figures 4.4* and *4.5*. When in operation, the whole unit is enclosed in a large, steel shell, as in *Figure 4.4*; this shell is filled with a dry gas under pressure to reduce the possibility of arcs, sparks, or corona discharges. The area where materials are irradiated with the electrons is usually in an adjacent room, as shown in *Figure B.4*.

An alternative source of high voltage is the Van de Graaff generator often used for the generation of high energy electrons. As shown in *Figure B.5*, a moving cloth belt carries the charge from a transformer–rectifier, typically supplying 20 000 V, up into a high-voltage cap at the top of the unit, which can reach a potential of 10 000 000 V. Elaborate electrical insulation is required and, once again, the whole unit is enclosed in a gas-filled pressurized shell.

A yet different type of linear accelerator is the resonant cavity or resonant transformer type, also called a 'linac'; this uses a series of accelerating units, as shown in *Figure B.6.* As the electrons drift from one section to the next, a pulse of radio-frequency is put down the tube so that it keeps in step with the electrons, continuously accelerating them. Energies of over 1000 MeV are possible in such units, which can also be used for protons and other particles. As the particles become accelerated and move more rapidly, the sections of the drift tube must be made successively longer, as shown in *Figure B.6.*

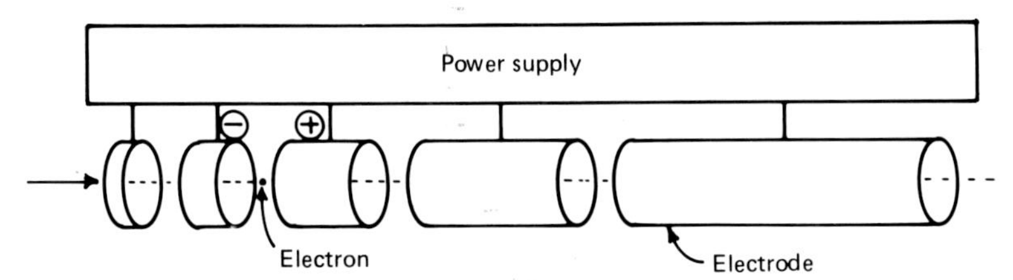

Figure B.6 The drift tube of a resonant-cavity linear accelerator; as the electron passes through an electrode, the voltage is switched to the next pair of electrodes

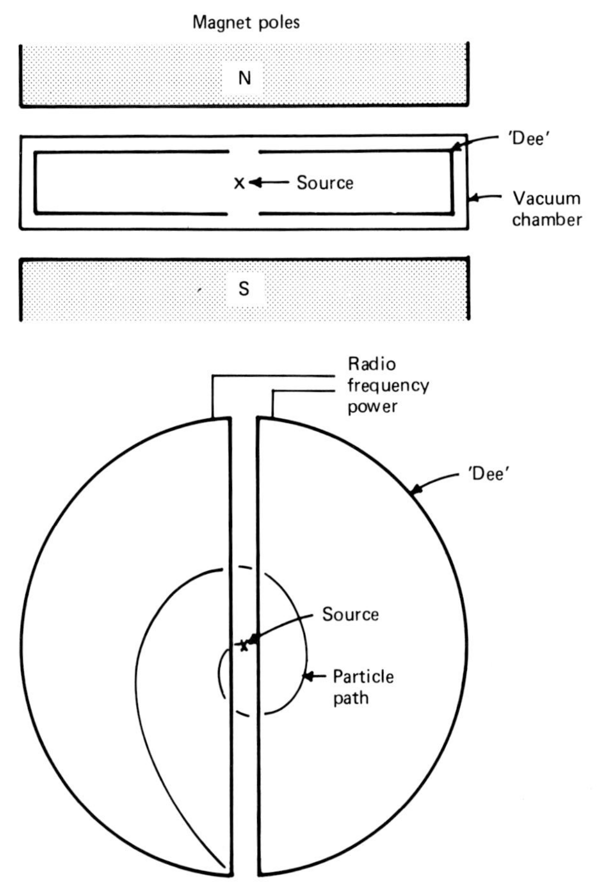

Figure B.7 Diagram of a cyclotron used to accelerate nuclear particles

Cyclotrons and related units, such as synchrotrons, cosmotrons, and the like, can be designed to accelerate electrons as well as positively charged particles, although they are usually used only for the latter. Here, two flat, D-shaped metal cylinders, called 'dees', are placed in a magnetic field, as in *Figure B.7.* Particles emitted from a source at the center follow a spiral path and are accelerated each time they cross from one dee to the other, because a high-frequency alternating potential applied to the dees keeps in step with the rotations.

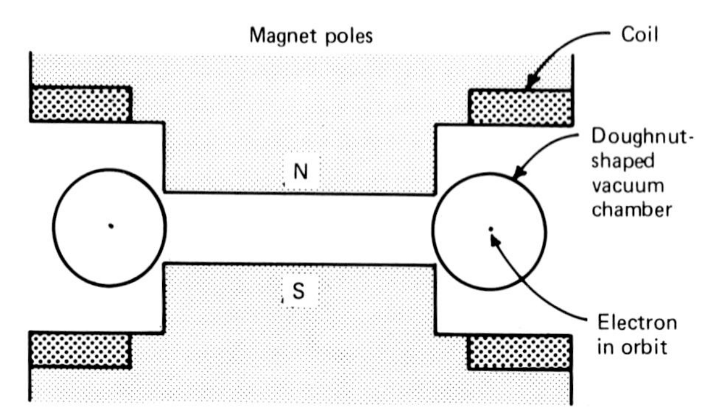

Figure B.8 Diagram of a betatron used to accelerate electrons

At high energies, the relativistic increase in mass of the electrons limits the usefulness of cyclotron-type machines; here betatrons can be used. These function like a transformer, with the electrons traveling in a doughnut-shaped tube that surrounds a large electromagnet activated with alternating current, as shown in *Figure B.8*; the electrons make many orbits during a single magnetic-field change. Energies of several hundred MeV can be reached.

The Decay of Radioactivity

When a material undergoes radioactive disintegration, the decay of this activity is best described by a half-life, usually written $t_{1/2}$. If the activity, in disintegrations per unit time or in percent of the activity remaining, is plotted against time, t, a curve such as that of *Figure B.9* is obtained. If the time at which the remaining activity has fallen to one-half, one-quarter, one-eighth, and so on is checked, one finds equal time intervals, as shown in the figure.

The equation which applies to such a process is:

$$dN = -\lambda N dt \quad \text{(B.10)}$$

where N is the total number of atoms, of which dN decay in a time interval dt, and λ is the decay or disintegration constant.

Integrating gives N_t, the amount remaining after time t, as:

$$N_t = N e^{-\lambda t} \quad \text{(B.11)}$$

and taking logarithms gives:

$$\log (N/N_t) = 0.434\lambda t \quad \text{(B.12)}$$

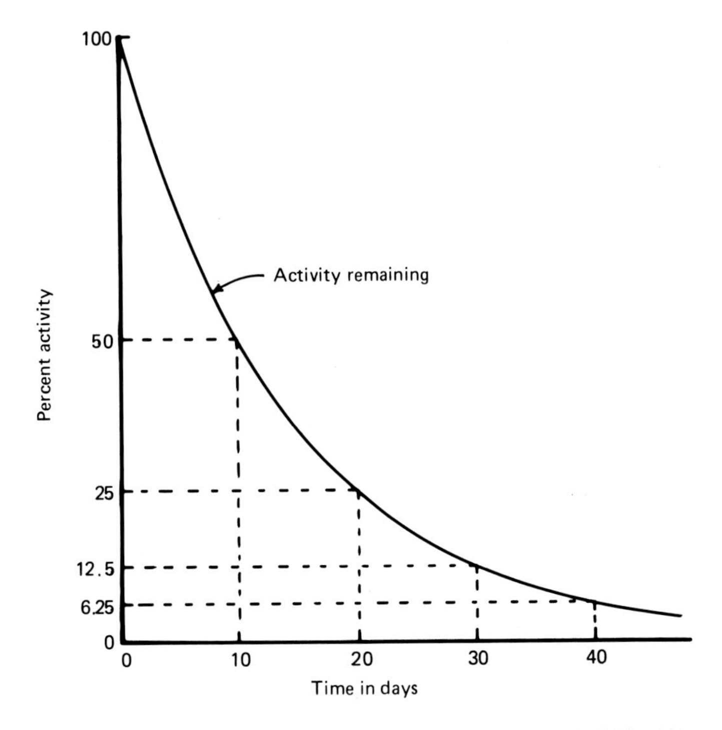

Figure B.9 The decay of radioactivity for a material having a half-life of 10 days, plotted with linear axes

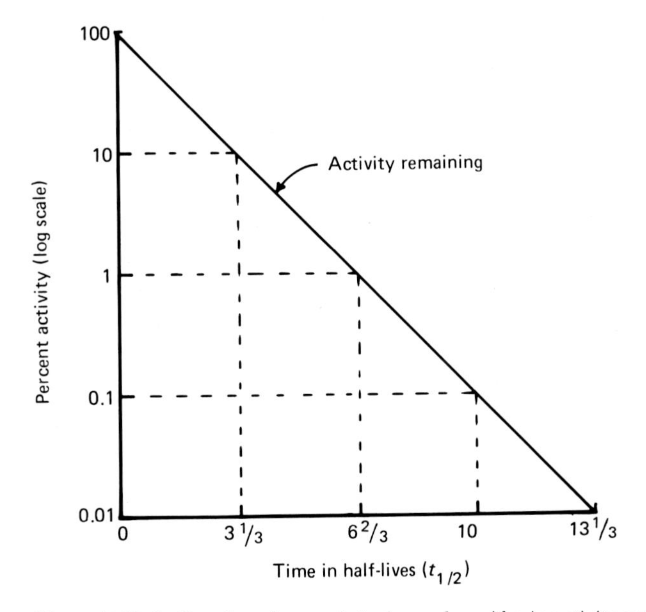

Figure B.10 Radioactive decay plotted on a logarithmic activity axis and using the half-life for the time axis. All materials follow the same line

When half the activity has disappeared, $N_t = N/2$, and $t = t_{1/2}$, the half-life. Inserting these values into equation (B.12) gives the half-life in terms of the decay constant λ:

$$t_{1/2} = 0.693/\lambda \qquad \text{(B.13)}$$

Equation (B.12) can be plotted in the form:

$$\log N_t = -0.434\lambda t + \log N \qquad \text{(B.14)}$$

to give a straight line, as in *Figure B.10*. By merely altering the horizontal scale so that the time units are replaced by units of the half-life, this straight-line graph now applies to the decay of any material.

The Detection of Radiation

The most commonly used radiation detection instrument is the Geiger or Geiger–Müller counter, which is one of a group of ionization counters. Other members of this group include proportional counters and ionization chambers. The Geiger counter usually consists of a chamber filled with gas at a low pressure; the chamber is equipped with an end or side window which is exposed to the radiation, as shown in *Figure B.11*. A central wire is charged to a high voltage by a battery, and there is electronic circuitry for measuring the current flow and displaying it, usually on a needle-type meter. When an ionizing particle, be it an electron, neutron, X-ray, or gamma ray, passes through the window and ionizes the gas, an electrical discharge occurs and an electric current flows between the central wire and the wall of the chamber, producing an indication on the metering system.

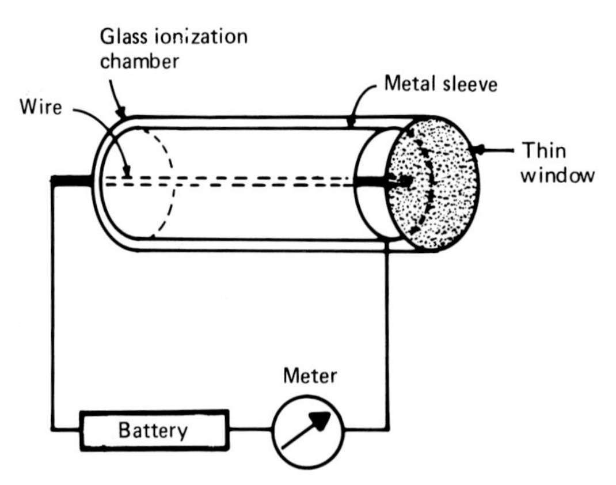

Figure B.11 Diagram of a Geiger counter

Geiger instruments are available as rugged, portable, relatively low-priced (a few hundred dollars) 'survey meters', such as that of *Figure B.12*, usually calibrated in counts or disintegrations per minute and mR or mrad per h; some also make a noise, a 'chirp', in a small speaker or earphones to indicate a response audibly.

Figure B.12 A battery-operated Geiger counter radiation survey-meter

Survey meters are used primarily for beta rays (electrons) and gamma rays; by placing a metal shield over the window, the beta rays are absorbed and then only gamma rays register. Heavy particles, such as protons and alpha particles, are usually completely absorbed by the window and special techniques are required for counting them. Neutrons are very inefficiently counted by such a system; if some boron-10 is present, however, the reaction:

$$ {}^{1}_{0}\mathrm{n} + {}^{10}_{5}\mathrm{B} \rightarrow {}^{7}_{3}\mathrm{Li} + {}^{4}_{2}\mathrm{He} \quad \text{(B.15)} $$

produces alpha particles, ${}^{4}_{2}\mathrm{He}$, which can then be counted.

Photographic film is used, particularly for individuals working with X-rays, in the form of small film-badges worn on the body. The amount of radiation is indicated by the amount of darkening produced on development of the photographic film. This test can also be used to provide a qualitative check of a gemstone or a group of gemstones for radioactivity by taking an 'autoradiograph'. The stone is merely laid for one to several days on top of the wrapped photographic film or plate (a self-developing film, such as 'Polaroid', is very convenient). On development, the radioactive stones will have taken their own photographs. An example is shown in *Figure 4.8*. A very intensely radioactive radium-treated diamond may take its own picture in just a few minutes[9]; such a stone should be kept in a lead container, as in *Figure 4.9*. Gamma rays by themselves have only a weak effect on a photographic film.

Table B.3 Summary of radiation detectors (modified after Coombe[3])

Type of counter	*Characteristics*
Geiger Geiger–Müller	Efficient for alpha- and beta-particles; not good for high counting rates
Proportional	Efficient for alpha- and beta-particles; good for high counting rates; suitable in modified form for neutrons
Ionization	Efficient for alpha particles, suitable for beta particles and gamma rays; good for high counting rates
Scintillation	Efficient for alpha- and beta-particles and gamma rays by choice of construction; good for high counting rates
Semiconductor	Efficient for all particles and rays

There are also pocket dosimeters of the electrometer type and a variety of other radiation-detector systems, including scintillation crystal-detectors as well as semiconductor crystal-detectors, which are not relevant in the present context. A summary of the characteristics of the different types of radiation detectors is given in *Table B.3*.

References

1. S. Glasstone, *Sourcebook on Atomic Energy,* 3rd Edn, Van Nostrand, Princeton, NJ (1967)
2. R. E. Lapp and H. L. Andrews, *Nuclear Radiation Physics,* 4th Edn, Prentice Hall, Englewood Cliffs, NJ (1972)
3. R. A. Coombe, *An Introduction to Radioactivity for Engineers,* Macmillan, London (1968)
4. A. Charlesby (Ed.), *Radiation Sources,* Macmillan, New York (1964)
5. J. Shapiro, *Radiation Protection: A Guide for Scientists and Physicians,* 2nd Edn, Harvard University Press, Cambridge, MA (1981)
6. K. Z. Morgan and J. E. Turner (Ed.), *Principles of Radiation Protection,* Wiley, New York (1967)
7. N. I. Sax, *Dangerous Properties of Industrial Materials,* 5th Edn, Van Nostrand Reinhold, New York (1975)
8. D. S. Billington and J. H. Crawford, Jr, *Radiation Damage in Solids,* Princeton University Press, Princeton, NJ (1961)
9. R. Crowningshield, Radium-treated diamond, *Gems Gemol.,* **12,** 304 (1968)

Appendix C

Color

Many of the treatments discussed in this book involve changes in color. To understand these changes and to predict what other changes may be possible, it is helpful to understand the causes of color in gemstones. The author's book *The Physics and Chemistry of Color*[1] gives all the causes of color that occur in nature or in man-made products. It describes the science and technology behind 15 types of selective emission or absorption of visible radiation, which produces the appearance of color when perceived by the eye. This book is written in such a way that it can also be read by the non-technical person, all the technical concepts being covered along the way.

Table C.1 The fifteen causes of color

VIBRATIONS AND SIMPLE EXCITATIONS

(1) *Vibrations and rotations*: water, ice, iodine, blue gas flame
(2) *Gas excitations*: vapor lamps, lightning, auroras
(3) *Incandescence*: flames, lamps, carbon arc, limelight

TRANSITIONS INVOLVING LIGAND-FIELD EFFECTS

(4) *Transition metal impurities*: ruby, emerald, aquamarine, jade, citrine
(5) *Transition metal compounds*: turquoise, rhodochrosite, malachite

TRANSITIONS BETWEEN MOLECULAR ORBITALS

(6) *Charge transfer*: blue sapphire and tourmaline, lapis lazuli
(7) *Organic compounds*: most dyes, amber, coral

TRANSITIONS INVOLVING ENERGY BANDS

(8) *Metals*: copper, silver, gold, brass
(9) *Pure semiconductors*: galena, cinnabar, diamond
(10) *Doped semiconductors*: blue and yellow diamond
(11) *Color centers*: amethyst, smoky quartz, Maxixe beryl

GEOMETRICAL AND PHYSICAL OPTICS

(12) *Dispersive refraction*: 'fire' in gemstones, rainbow
(13) *Scattering*: moonstone, blue sky, red sunset
(14) *Interference*: iris quartz, iridescent chalcopyrite, oil slick on water, soap bubbles
(15) *Diffraction*: opal, iris agate, diffraction gratings

No less than 13 of the 15 types of color outlined in *Table C.1* apply to gem, mineral, and jewelry substances, and at least 11 to the gemstones and treatments discussed in this book. For full details and further references the book[1] should be consulted. There are also more limited treatments which may be useful[2–4].

Light and Energy Considerations

For the pure colors of the spectrum, the visually observed color depends on the wavelength of the light; thus 580 nm corresponds to yellow. An energy can also be assigned to each wavelength in spectrally pure colors, as is done in *Figure C.1*, in terms of the electron volt described in Chapter 4, p. 47; yellow is about 2.15 eV. A well-known analogy shows that energy increases in going through the spectrum

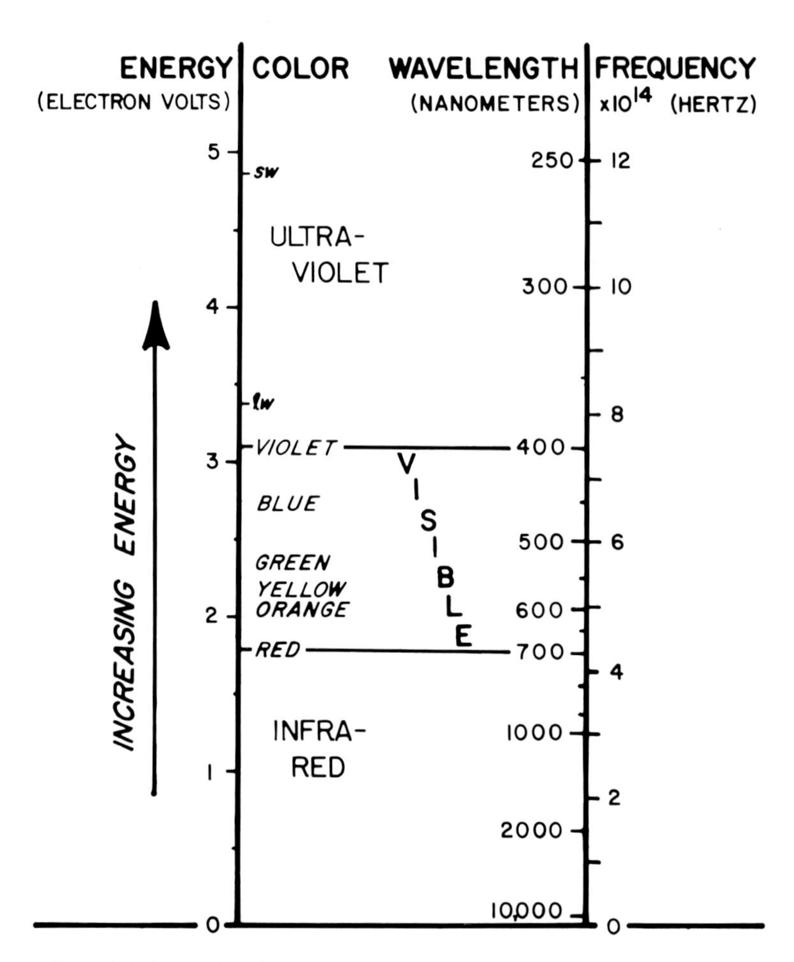

Figure C.1 Increasing the temperature increases the energy and decreases the wavelength of the emitted light; sw is short-wave ultraviolet at 254 nm, lw is long-wave ultraviolet at 366 nm. From K. Nassau, *The Physics and Chemistry of Color*, John Wiley and Sons, New York (1983)

Table C.2 Various ways of specifying the visible and near-visible spectrum

Color[a]	*Wavelength description;* λ			*Frequency,* ν (Hz $\times 10^{-14}$)	*Wavenumber,* $\bar{\nu}$ (cm^{-1})	*Energy, E* (eV)
	Ångstrom (Å)	*Nanometer* (nm)	*Micrometer* (μm)			
Infrared (far)	300 000	30 000	30	0.1	333	0.041
Infrared (near)	10 000	1 000	1.00	3.00	10 000	1.24
Red (limit)	7 000	700	0.70	4.29	14 300	1.77
Red	6 500	650	0.65	4.62	15 400	1.91
Orange	6 000	600	0.60	5.00	16 700	2.06
Yellow	5 800	580	0.58	5.16	17 240	2.14
Green	5 500	550	0.55	5.45	18 200	2.25
Green	5 000	500	0.50	5.99	20 000	2.48
Blue	4 500	450	0.45	6.66	22 200	2.75
Violet	4 000	400	0.40	7.50	25 000	3.10
Ultraviolet (lw)	3 660	366	0.366	8.19	27 300	3.39
Ultraviolet (sw)	2 537	254	0.254	11.8	39 400	4.89

[a] Typical values only; lw stands for long-wave, sw for short-wave.

from red to blue; when an object is heated it becomes first 'red-hot', then orange-, yellow- and white-hot; if hot enough, as in the very hottest of stars, it becomes 'blue-hot'. At even higher energies the ultraviolet region is reached, as shown in *Figure C.1*.

Some of the many additional ways of specifying the colors of the visible and near-visible spectrum are given in *Table C.2,* while some of the equations used for conversions are given below.

Definitions and relationships connected with light and color

(a) Velocity of light, c, $= 2.9979 \times 10^{10}$ cm/sec in vacuum.

(b) Wavelength λ is the distance between two repeating parts of the light wave in centimeters, cm, micrometers, μm, nanometers, nm, angstroms, Å, etc.

(c) Wavenumber $\bar{\nu}$ is the number of waves in a length of 1 cm:

$$\bar{\nu} = \frac{1}{\lambda(\text{cm})} = \frac{10\,000}{\lambda(\mu\text{m})} = \frac{10\,000\,000}{\lambda(\text{nm})} = \frac{100\,000\,000}{\lambda(\text{Å})}$$

(d) Frequency ν is the number of waves of light passing a point in one second, previously cycles per sec, now hertz, Hz:

$$\nu = \bar{\nu} \times 3 \times 10^{10}\ \text{Hz}$$

(e) Energy, E

$$\begin{aligned} E &= \bar{\nu} \times 1.9865 \times 10^{-16}\ \text{ergs} \\ &= \bar{\nu} \times 1.9865 \times 10^{-23}\ \text{joules} \\ &= \bar{\nu} \times 1.2399 \times 10^{-4}\ \text{electron volts} \\ &= \bar{\nu} \times 4.7478 \times 10^{-24}\ \text{calories} \\ &= \bar{\nu} \times 2.86\ \text{calories per gram molecule} \end{aligned}$$

(f) Useful conversion between wavelength and energy:

$$(\text{wavelength}, \lambda, \text{in nm}) \times (\text{energy}, E, \text{in eV}) = 1239.9$$

References

1. K. Nassau, *The Physics and Chemistry of Color: The Fifteen Causes of Color,* Wiley, New York (1983)
2. K. Nassau, The origins of color in minerals and gems, *Lap. J.,* **29,** 920, 1060, 1250, 1521 (1975); also *Gems Gemol.,* **14,** 354 (1974–1975); **15,** 2, 34 (1975)
3. K. Nassau, The origins of color in minerals, *Amer. Min.,* **63,** 219 (1978)
4. K. Nassau, The causes of color, *Scientific American,* **243,** 106, and cover (Oct 1980)

Appendix D

Purveyors of Supplies and Services

The listings of purveyors, primarily in the US, are given under the headings:

Ceramics and heating elements for building furnaces
Chemicals
Furnaces, general, and laboratory supplies
Furnaces, special and high temperature
Gases
Gemology courses
Gemology and related periodicals
Gemstone identification and certification
Gemstone treatment: heating
Gemstone treatment: irradiation
Gemstone treatment: lasering
Radiation equipment
Temperature control and measurement

These listings are intended for general information only. They are given in alphabetical order within each group and are not intended to be complete; neither is any specific recommendation implied. A number of these suppliers have regional offices throughout the US; the head-office location is usually listed.

Ceramics and Heating Elements for Building Furnaces

Carborundum Co, Globar Division (heating elements), PO Box 339, Niagara Falls, NY

Coors Porcelain Co, 600 Ninth St, Golden, CO 80401

Kanthal Furnace Products Corp (heating elements), Wooster St, Bethel, CT 06801

McDanel Refractory Porcelain Co, 510 Ninth Ave, Beaver Falls, PA 15010

National Ceramic Co, PO Box 280, Trenton, NJ 08602

Norton Co, Industrial Ceramics Div, One New Bond St, Worcester, MA 01606

Chemicals

Aldridge Chemical Co, PO Box 355, Milwaukee, WI 53201

Alfa Products, Thiokol/Ventron Div, 152 Andover St, Danvers, MA 01923

Atomergic Chemetals Corp, 100 Fairchild Ave, Plainview, NY 11803

J. T. Baker Chemical Co, 222 Red School Lane, Phillipsburgh, NJ 08865

Sigma Chemical Co, PO Box 14508, St Louis, MO 63178

Furnaces, General, and Laboratory Supplies

This includes crucibles, crucible tongs, safety goggles, etc.

American Scientific Products, 1430 Waukegan Rd, McGaw Park, IL 60085

Fisher Scientific Co, 711 Forbes Ave, Pittsburgh, PA 15219

Lab Safety Supply, Division of Science Related Materials, PO Box 1368, Janesville, WI 53574

SGA Scientific Inc, 735 Broad St, Bloomfield, NJ 07003

Thermolyne, Division of Sybron Corp, 2555 Kerper Blvd, Dubuque, IA 52001

Furnaces, Special and High Temperature

Astro Industries, Inc, 606 Olive St, Santa Barbara, CA 93101

Blue M, A Unit of General Signal, 138th and Chatham, Blue Island, IL 60406

C M Manufacturing and Machine Co, Inc, 103 Dewey St, Bloomfield, NJ 07003

Centorr Associates, Inc, Rte 28, Suncook, NH 03275

Harper Electric Furnace Corp, 303 W. Drullard Ave, Lancaster, NY 14086

Le Mont Scientific, 2011 Pine Hall Dr, State College, PA 16801

Lindberg, A Unit of General Signal, 304 Hart St, Watertown, WI 53094

Gases

Airco Industrial Gases, 575 Mountain Ave, Murray Hill, NJ 07974

Linde Division of Union Carbide Corp, 100 Davidson Ave, Somerset, NJ 08873

Matheson, 932 Paterson Plank Rd, East Rutherford, NJ 07073

Scientific Gas Products, Inc, 2330 Hamilton Blvd, South Plainfield, NJ 07080

Gemology Courses

Organizations offering correspondence courses are marked with *

Gemological Institute of America, 580 Fifth Ave, New York, NY 10036

Gemological Institute of America*, 1660 Stewart St, Santa Monica, CA 90404

Gemmological Assn of Great Britain*, St Dunstan's House, Carey Lane, London EC2V 8AB, England

Gemology and Related Periodicals

Primarily Gemology

Gems and Gemology
Gemological Institute of America, PO Box 2110, Santa Monica, CA 90406

The Journal of Gemmology
Gemmological Assn of Great Britain, St Dunstan's House, Carey Lane, London EC2V 8AB, England

Related Subjects

Jeweler/Gem Business
5870 Hunters Lane, El Sobrante, CA 94803

Jewelers' Circular-Keystone
Chilton Way, Radnor, PA 19089

Lapidary Journal
PO Box 80937, San Diego, CA 92138

Modern Jeweler
PO Box 2939, Shawnee Mission, KA 66201

National Jeweler
1515 Broadway, New York, NY 10036

Gemstone Identification and Certification

American Gemological Laboratories, 645 Fifth Ave, New York, NY 10022

Gem Trade Laboratory, Inc (a subsidiary of GIA), 580 Fifth Ave, New York, NY 10036

Gem Trade Laboratory, Inc (a subsidiary of GIA), 1660 Stewart Street, Santa Monica, CA 90404

Gem Trade Laboratory, Inc (a subsidiary of GIA), 606 South Olive St, Los Angeles, CA 90014

International Gemmological Institute, 20 W 47th St, New York, NY 10036

Gemstone Treatment: Heating

Primarily corundum (sapphires, ruby)

ICT Inc., L. P. Kelley, 1330 Industrial Drive, Shelby, MI 49455

Lambertville Ceramic Co, B. Cass, 245 Main St, Lambertville, NJ 08530

Salie, Y. M., 2 Deseret Drive, East Brunswick, NJ 08816

Primarily materials other than corundum

International Lapidaries, J. Call, PO Box 359, Boynton Beach, FL 33429

Plumbago Mining Corp, D. McCrillis, PO Box 447, Rumford, ME 04275

Gemstone Treatment: Irradiation

Borden, J., PO Box 99387, San Diego, CA 92109

CN & L Investment Corp., G.P. Drake, PO Box 2612, California City, CA 93505

IRT Corp, 3030 Callan Rd, San Diego, CA 92121

Nu Age Products, Inc, 1244 River Rd, Hyde Park, MA 02136

Nuclear Theory and Technologies, C. E. Ashbaugh, 6146 Jumille Ave., Woodland Hills, CA 91367

Process Technology, Inc, R. Buckley, North Airport Rd, West Memphis, AK 72301

Radiation Sterilizers, Inc, 3000 Sand Hill Rd, Menlo Park, CA 94025

Radiation Technology, Inc, M. Welt, Lake Denmark Rd, Rockaway, NJ 07866

Gemstone Treatment: Lasering

Irving Newman and Son, 580 5th Ave, New York, NY 10036

Perlman Brothers, 71 W 47th St, New York, NY 10036

Romella Lasing, 36 W 47th St, New York, NY 10036

Radiation Equipment

Beta Analytical, Inc, 136 Bradford Ave, Pittsburgh, PA 15205

E G and G Ortec, 100 Midland Rd, Oak Ridge, TN 37830

National Electrostatics Corp, Box 310C, Graber Rd, Middletown, WI 53562

New England Nuclear, 549 Albany St, Boston, MA 02118

Temperature Control and Measurement

See also: Furnaces, General, and Laboratory Supplies

Cole Parmer Instrument Co, 7425 N. Oak Park Ave, Chicago, IL 60648

Leeds and Northrup, A Unit of General Signal, Sumneytown Pike, North Wales, PA 19454

Omega Engineering, Inc, 1 Omega Drive, Stamford, CT 06907

Perkin-Elmer Corp, 2000 York Rd, Oak Brook, IL 60521

Index

The main treatment section for a specific gemstone is shown in bold numbers; color plates are listed by their Roman numbers.